Philosophy of Engineering and Technology

Volume 35

The *Philosophy of Engineering and Technology* book series provides the multifaceted and rapidly growing discipline of philosophy of technology with a central overarching and integrative platform. Specifically it publishes edited volumes and monographs in: the phenomenology, anthropology and socio-politics of technology and engineering the emergent fields of the ontology and epistemology of artifacts, design, knowledge bases, and instrumentation engineering ethics and the ethics of specific technologies ranging from nuclear technologies to the converging nano-, bio-, information and cognitive technologies written from philosophical and practitioners' perspectives and authored by philosophers and practitioners. The series also welcomes proposals that bring these fields together or advance philosophy of engineering and technology in other integrative ways. Proposals should include: A short synopsis of the work or the introduction chapter. The proposed Table of Contents. The CV of the lead author(s). If available: one sample chapter. We aim to make a first decision within 1 month of submission. In case of a positive first deci+sion the work will be provisionally contracted: the final decision about publication will depend upon the result of the anonymous peer review of the complete manuscript. We aim to have the complete work peer-reviewed within 3 months of submission. The series discourages the submission of manuscripts that contain reprints of previous published material and/or manuscripts that are below 150 pages/75,000 words. For inquiries and submission of proposals authors can contact the editor-in-chief Pieter Vermaas via: p.e.vermaas@tudelft.nl, or contact one of the associate editors.

More information about this series at http://www.springer.com/series/8657

Simona Chiodo • Viola Schiaffonati
Editors

Italian Philosophy of Technology

Socio-Cultural, Legal, Scientific and Aesthetic Perspectives on Technology

Springer

Editors
Simona Chiodo
Department of Architecture
and Urban Studies
Politecnico di Milano
Milano, Italy

Viola Schiaffonati
Department of Electronics,
Information and Bioengineering
Politecnico di Milano
Milano, Italy

ISSN 1879-7202 ISSN 1879-7210 (electronic)
Philosophy of Engineering and Technology
ISBN 978-3-030-54521-5 ISBN 978-3-030-54522-2 (eBook)
https://doi.org/10.1007/978-3-030-54522-2

This Springer imprint is published by the registered company Springer Nature Switzerland AG
The registered company address is: Gewerbestrasse 11, 6330 Cham, Switzerland

Contents

Chapter 1
The Italian Philosophy of Technology

Simona Chiodo and Viola Schiaffonati

Abstract The introduction presents this volume collecting the contributions from Italian philosophers who have produced reflections on technology over recent decades. This volume represents the opportunity to present to an international audience the work of scholars who, to date, have had mainly a national resonance (with few exceptions). This is the first volume of its kind published in English, and its purpose is to offer an idea of the multifaceted development of the Italian philosophy of technology over recent decades, especially to the non-Italian reader.

Keywords Humanities philosophy of technology · Analytic philosophy of technology · Technique and technology · Culture · Society and politics · Philosophy of law · Philosophy of science · Aesthetics

This volume presents contributions from Italian philosophers who have produced reflections on technology over recent decades. Twelve out of the fifteen contributions have been explicitly written for this volume, and represent the authors' main views on technology. Although some of the works of these authors were already written in English and have been published in international venues, this is not true for all of them. Thus, this volume represents the opportunity to present to an international audience the work of scholars who, to date, have had mainly a national resonance (with few exceptions). This is the first volume of its kind published in English, and its purpose is to offer an idea of the multifaceted development of the Italian philosophy of technology over recent decades, especially to the non-Italian reader.

S. Chiodo (✉)
Department of Architecture and Urban Studies, Politecnico di Milano, Milan, Italy
e-mail: simona.chiodo@polimi.it; viola.schiaffonati@polimi.it

V. Schiaffonati
Department of Electronics, Information and Bioengineering, Politecnico di Milano, Milan, Italy

S. Chiodo, V. Schiaffonati (eds.), *Italian Philosophy of Technology*, Philosophy of Engineering and Technology 35,
https://doi.org/10.1007/978-3-030-54522-2_1

1.1 Humanities Philosophy of Technology and Beyond

According to Carl Mitcham's definition (Mitcham 1994), we may broadly say that most of the Italian philosophy of technology is an articulation of what he calls *humanities philosophy of technology*, in that it is founded on classical philosophical methodologies, concepts, and authors belonging to the historical tradition of the humanities. This is true for most of the authors of this volume, who are philosophers by training and make use of concepts and authors coming from the history of philosophy. Consequently, in the Italian speaking community focused on the analysis of technology, limited attention has been devoted to the very practice of technology and to what is referred to as analytic philosophy of technology (Franssen et al. 2018).

The Italian philosophy of technology is now partly changing and, as this volume shows, the empirical turn in the European and North-American philosophy of technology has influenced the Italian community as well (Franssen et al. 2016). On the one hand, the contributions of the oldest generation of scholars are mainly (but not exclusively) characterised by the typical traits of the humanities philosophy of technology. On the other hand, the contributions of the latest generation combine the humanities philosophy of technology with an empirically informed philosophy of technology attentive to the practice of engineering. As for the oldest generation, the tools are mainly drawn from metaphysics, and the inspiring authors are the classical ones, such as Kant and Heidegger. Besides, the reference to the historical framework is central, since it is thought to be indispensable to fully understand the meaning of the various phenomena considered. From this perspective, we may say that the meaning of a single phenomenon comes from the wider context of both diachronic (i.e., historical) and synchronic (i.e., contemporary) cultural phenomena, more than from the single phenomenon itself. As for the last generation, the tools are mainly drawn from philosophy of science and philosophy of law, and the inspiring authors are more often contemporary analytic philosophers. What is interesting, and even promising, is that, since Italian philosophers usually have a strong historical background by training, the more recent philosophy of technology is experimenting an interesting mix that includes both a deep awareness of the complex framework and a higher attention to the very practice of technology.

More specifically, what is promising is the approach according to which the philosophy of technology starts from an in-depth knowledge of the practice of technology and engineering, and then integrates it with the exercise of what Umberto Eco called the "vision of totality" (Eco 1968). In our case, this means trying to provide a doubly far-sighted perspective: in order to grasp the philosophical meaning of a certain technology, philosophers of technology should be able to make reference both to its past (for instance, to the various and complex reasons of its choice among its possible alternatives) and to its future (for instance, to the possible scenarios it is likely to have impact on, or even cause). We think that this is precisely what the extreme sophistication of contemporary technologies demands. Paradoxically enough, we live in a time when the classical meaning of philosophers as far-sighted humanists who can have the "vision of totality" is more useful than ever. Dealing with extremely powerful, as well as quickly changing, technologies requires a capacity for vision that largely exceeds the present circumstances, as well as a capacity to anticipate the future. In doing this, both the awareness of the past and the training to grasp the complex relationships between a certain technological artefact and its cultural and social framework play an essential role.

1.2 Technique and Technology

In the analysis of the traits of the Italian philosophy of technology, another symptom of its specificity is the extensive use of the Italian word *tecnica* (technique) instead of the Italian word *tecnologia* (technology). To reflect on the meaning of the distinction between *tecnica* and *tecnologia* within the Italian humanities philosophy of technology is important to highlight one interesting aspect. While the word *tecnologia* in Italian usually means the complex artefact whose condition of existence is the joined work of both science and technology, the word *tecnica* usually means something simpler. It means a simpler technical artefact, which does not necessarily involve any scientific methodology in its design and development. It can even express a simple way of doing something. For instance, we can say that we are using a *tecnica* when we are making pasta by hand following precise movements and rhythms. This use shows that a part of the Italian philosophy of technology of the second half of the last century focused on a notion of artefact that does not necessarily involve a complex methodology such as the one developed in the sciences. Rather, it mainly focused on the anthropological precondition, as it were, of *tecnologia* itself: in other words, on the human aptitude for using both human and artificial resources to change the natural world. Accordingly, this approach focuses more on the reasons why a human being builds a certain kind of artefact in a certain kind of way than on the functioning of the artefact itself and its use in the everyday practice of technology and engineering.

This kind of philosophical investigation largely requires metaphysical and historical tools. Accordingly, the typical questions are the following: are human beings essentially technological, namely, are they characterised by an essential aptitude for technology? If this is not the case, how can we understand technology? Vice versa, if this is the case, how can we understand human history? And how can we understand human destiny? According to this philosophical approach, the very meaning of a technological artefact is given more by its metaphysical condition of existence than by its actual functioning, and the expertise required to investigate it is metaphysical.

This approach is partly changing today. As already mentioned, the most recent contributions of the Italian philosophy of technology integrate the metaphysical tradition with the empirical turn characterising the analytic philosophy of technology. Accordingly, the word *tecnica* is substituted with the word *tecnologia* to give reason, also from a terminological point of view, for this change of perspective. In these contributions the object of study requires scientific and technological competences, as well as metaphysical and historical ones. This means that the very practice of the philosophy of technology is moving away from a mainly solitary reflection, in order to reach a more collaborative work including interactions with scientists and technologists (as per our own experience of philosophers working in a technical university).

1.3 The Issues at Stake

Italian philosophy is known to be particularly focused on issues related to society and politics. This is also true in the case of the philosophy of technology. Indeed, many of the authors collected in this volume reflect, whether directly or indirectly, on the social context from which technology arises, and especially on which technology impacts. In some cases, these reflections are oriented towards socio-cultural issues, such as the specificity of Western culture related to the development of technology, the relationship between the notions of human beings and technology, as well as the peculiarity of the technical artefact among other kinds of artefacts. In some other cases, the authors' reflections are oriented towards socio-political issues, such as the impact of technology on morality, ethics, and the law, the relationship between economy and technology, as well as how technology is going to reshape the relationships between both citizens and nations.

In an attempt to find a common thread among the issues analysed in the chapters of this volume, we could say that it has to do with a kind of crisis. More precisely, the crisis of the subject as this notion has been ontologically conceived in Western philosophy. Many contributions reflect on this issue in pessimistic, or at least critical, terms. The subject, as Kant founded it, is losing autonomy (chapters by Pagallo and Artosi). Thus, a progressively less ruling and controlling subject requires a new ethics (chapter by Fabris). Contemporary art also shows a crisis of both the arts themselves and the subject, who progressively moves her/his prerogatives from inside to outside (chapter by Costa). And, even if the terms are less pessimistic, what is at stake is an ontological revolution. In the case of the chapters by Floridi and Diodato, and to a certain extent also of those by Roversi, Datteri, Tamburrini and Altiero, Contissa, Godano and Sartor, this revolution characterises both technology and humankind, whereas in Magnani's chapter this revolution deals with morality. This ontological revolution is certainly a product of human nature itself, being technological in its essence, as is clear in the chapters by Ferraris, Severino, Sini and Montani.

1.4 The Structure and the Contents

This volume is organized into four different parts corresponding to the main contributions given by Italian scholars to the philosophy of technology over the last decades. As the first part shows, a major issue has to do with the relationship between technology on the one hand, and the cultural milieu from which technology arises, as well as on which technology impacts, on the other hand. The cultural milieu is often taken into account in socio-political terms. Not surprisingly, as emphasised in the second part, another important contribution to the reflection on technology is given by the Italian philosophy of law, which is generally characterised by a stronger interest in applied research, moving from a more metaphysical approach (often characterising the first part of the volume) to a more practical one. A significant aspect of the Italian philosophy of science also focuses on technology, as the third part of this volume shows, usually attesting a stronger connection to the empirical turn mentioned in Sect. 1.1. Finally, in the fourth and last part of the volume, Italian aesthetics shows a particularly strong interest in technology, both when conceived as philosophy of perception and when conceived as philosophy of art, also including a more general reflection on the aesthetic relationship between artefacts and human beings as their users.

1.4.1 Culture, Society and Politics

The first part gathers the contributions that can be collected under the very general heading *Culture, Society and Politics*, areas that more typically characterise the Italian philosophical tradition. This part starts with the article of a philosopher (Maurizio Ferraris) who comes from the hermeneutic tradition, on the one hand, and from the influence of French philosophy (especially of Derrida's work), on the other hand. Ferraris has developed an original philosophical proposal, as well as an original philosophical writing style, both by working on social ontology through a new form of realism and by discussing with authors coming from the Anglo-American tradition. He proposes to change the classical vision of the human being as "the animal endowed with language" with the vision according to which the human being is "the animal endowed with a stick". This definition implies that humans are essentially technological, and that technology is not their cause of alienation. More precisely, technology is what most completely and powerfully reveals what humans essentially are, i.e., beings unsatisfied with their own condition, and therefore trying to change it by using the technology of the stick in order to evolve. Accordingly, humans are essentially dependent on technology, which is what makes them able to capitalise strength, authority and memory.

Adriano Fabris's article also deals with the control of technology, by arguing for the necessity of a new ethics capable of overcoming the somehow rigid boundaries between the analytic and the continental traditions. The most important issue of this new ethics is the reflection on how to reframe the notion of responsibility, as well as on how to use it, and thus continue to distinguish humans from today's very sophisticated machines.

The first part also includes an article by Luciano Floridi, whose work in the philosophy of technology is very well-known outside of Italy. In his article, Floridi proposes his ontological interpretation of informational privacy. In particular,

Floridi argues for the moral and legal potentially advantageous outcomes of his novel conception of informational privacy. By founding his reflection on his own notion of infosphere, Floridi claims that the forces opposing the informational flow within the infosphere can be defined in terms of ontological friction: the more the ontological friction decreases, the more accessibility to personal data increases.

The first part ends with the contributions of two philosophers who have made notable impressions on the Italian philosophical tradition for several decades. They are both distinguished representatives of the so-called continental philosophy, and remarkably original in their philosophical paths, as also their writing styles show. The first one is Emanuele Severino, who passed away only some months after having delivered his original article for this volume. He strongly influenced Italian philosophy in general, and his article, which represents his particular style, can be better understood within his broader philosophical view. According to Severino, Western culture is characterised by a special form of nihilism, which conceives things as a denial of the being. This special form of nihilism is particularly strong in our era, where an increasingly powerful techno-science has finally become its own primary aim. Techno-science considers things as something to continuously produce and destroy. Accordingly, techno-science has moved from being a means to being the supreme planetary aim of any dominating structure (from capitalism in general to the most powerful states in particular), where techno-science is the *passe-partout* to attempt to rule the world. In this scenario, Severino's article reflects on the domination of technology, especially from a metaphysical point of view, by considering some of the major steps in the history of philosophy. He claims that the domination of technology will continue in the future, particularly in a world where some classical metaphysical issues, such as the notions of truth and reality, are getting lost.

The second one is Carlo Sini's article. It reflects Sini's background in the philosophical traditions of phenomenology and hermeneutics. Sini has developed an original philosophical stance as well, by working especially on the notion of writing, and in particular on the Greek alphabet, as a crucial foundation of the very Western way of thinking. Sini attributes to the notion of writing precisely a technological meaning, which is increasingly dominating our era. In his article, starting from Heidegger's reflection on technology, Sini argues that it is precisely the instrument, namely, the instrumental action, what determines the human identity. This instrumental action makes it possible for humans to conceive the world in which they act as object. By reflecting on the use of human language, he claims that also the very notion of culture is a special kind of machine, being artfully produced.

1.4.2 Philosophy of Law

The second part presents the contributions of the philosophy of law. Giuseppe Contissa, Francesco Godano and Giovanni Sartor provide the reader with an overview of the major contributions of the Italian legal philosophy to the emergence of information technologies between the 1960s and the 1970s. Four authors coming from four different philosophical backgrounds are taken into account: idealism (Vittorio Frosini), analytic philosophy combined with legal positivism (Mario Losano), the critique of legal positivism combined with ethical/axiological/theological ideas (Luigi Lombardi Vallauri) and the more concrete, policy-oriented and revolutionary approach of a practitioner (Renato Borruso).

Ugo Pagallo's article focuses on the critical issues related to the principle of privacy by design, according to which we try to embed data protection safeguards in ITC and other technologies. Whereas this attempt may be promising in data protection in CCTV systems, biometrics, social networks, smart environments and data loss prevention, it may be quite critical when it comes to facing complex ethical issues, from individual autonomy to the democratic rule of law, as well as to anti-paternalism. Indeed, automatic techniques determining people's behaviours risk to decrease free individual and political choices. Thus, Pagallo argues that the principle of privacy by design should, in any case, encourage people to change their conduct by strengthening their rights and broadening their range of choices, and that a promising strategy may be relying on "digital air-bags", i.e., devices, that, as friendly interfaces, reduce the impact of harm.

In order to explain the very notion of law, Corrado Roversi investigates its artefactual nature. Roversi, who comes from a philosophical background, introduces his own historical-intentional model of the artefactuality of legal institutions, according to which they are the outcome of a history of intentions, and shows how the law is embedded within the broader, distinctly human, practice of building artefactual objects endowed with symbolic meaning.

The second part ends with the work of an author whose background is rather particular: Alberto Artosi is a logician and a philosopher of science by training, as well as a scholar of Leibniz's philosophy. His philosophy of law is strongly influenced by his background, and in particular by the logic of moral theories, on the one hand, and by dialectical reasoning and arguing, on the other hand. In his article, Artosi focuses on technical normativity, i.e., normativity concerning artefacts and their features. He analyses a strikingly growing class of technical norms more or less explicitly directed at influencing human behaviours. What emerges from this analysis is a non-legal form of normativity involving the subject in a process of continuous observation, surveillance and direction of her/his conduct, whose final aim is her/his self-identification with certain imposed standards.

1.4.3 Philosophy of Science

The third part collects the contributions within the context of the philosophy of science. In their article, Guglielmo Tamburrini and Francesco Altiero examine machine intelligence games from a methodological perspective, focusing on their role as generators of research programmes. By reviewing some cases, they underline interesting advantages offered by the gamification of machine intelligence research. This process helps not only feed the pipeline between academic research and industry, but also motivate junior researchers, attract wider cohorts of science and engineering students, as well as raise public awareness and interest towards AI and robotics.

Robotics is also the subject of Edoardo Datteri's contribution, which discusses robots as experimental platforms for the study of animal and human behaviours. Datteri argues that robots can be used to promote mechanistic conceptions of animal behaviour, but also, and perhaps moreover, to test specific behavioural hypotheses. More specifically, he shows by means of examples how, in contemporary biorobotics, robots are expressly built in order to be involved in experiments aiming at studying animal and human behaviours.

The idea of knowledge as a duty is the core of Lorenzo Magnani's article, which builds on the reflections of a previous volume with the same title. Magnani's philosophical proposal is founded on two concepts in particular: respecting people as things and conceiving technological artefacts as moral mediators. As for the first issue, the author develops his reflection starting from the previous results of his analysis of both Kant's and Mill's ethics, and then by overcoming them through a new proposal: the attribution of moral values to things. As for the concept of moral mediator, Magnani reflects especially on the case in which a technological artefact acts as a moral mediator. This means that a technological artefact mediates objectively some moral effects out there in the world, and therefore gives us important moral knowledge by acting as an objectified moral device.

1.4.4 Aesthetics

The fourth and last part collects the contributions of aesthetics on technology. Mario Costa's article starts from the analysis of some contemporary works of art, which are to some extent the realization of his notion of technological sublime. His thesis is that, over the last two centuries, technology has somehow corrupted the essence of art, and has finally dissolved art itself into a form of technological sublime. This phenomenon means overcoming the notion of art as it has been traditionally conceived, and makes us react in a way that is analogous to the sublime, as has been defined by several philosophers.

Roberto Diodato's contribution focuses on the ontology of the virtual, which deserves to be carefully studied by taking into account its novel technological developments. The virtual, which results from the convergence of computer programming, on the one hand, and of human bodies augmented through technological prostheses, on the other hand, is conceived as a new entity, which we may define in terms of an interactive body-image.

Finally, Pietro Montani's contribution makes reference to Kant's philosophy in order to establish his own concept of techno-aesthetics. Of course, Kant's theoretical tools need to be re-thought to satisfactorily address the radically novel outcomes of technological innovation. Nevertheless, it is the Kantian notion of genius what founds, indeed, the very conditions to welcome the infinite innovativeness characterising the present (as well as the future) articulations of technological creativity.

References

Eco, U. (1968). *La struttura assente: introduzione alla ricerca semiologica*. Milano: Bompiani.
Franssen, M., Vermaas, P. E., Kroes, P., & Meijers, A. (Eds.) (2016). *Philosophy and technology after the empirical turn*. Cham: Springer.
Franssen, M., Lokhorst, G. J., & Van de Poel, I. (2018). Philosophy of technology. In E. N. Zalta (Ed.), *Stanford encyclopaedia of philosophy (Fall 2018 Edition)*. Stanford, CA: Stanford University.
Mitcham, C. (1994). *Thinking through technology: The path between engineering and philosophy*. Chicago: University of Chicago Press.

Part I
Culture, Society and Politics

Chapter 2
Oedipus' Stick

Maurizio Ferraris

Abstract As children we are often told about the riddle of the sphinx. What animal has only one voice (i.e. one essence, which consists in being endowed with language) but walks in three different ways: in the morning on four legs, at noon on two, and in the evening on three? That animal, as we all know, is the human being, but the answer suggests that being endowed with language and *hands* is not enough: it is not the human essence, its fundamental *eidos*. The technical prosthesis (the stick) thus falls within the very definition of the human animal, something that does not apply to other animals. This circumstance shows the inner link between technology and human nature.

The human being is not a very intelligent animal. This explains many things, except one, namely why it has been called a "rational animal". Perhaps because it speaks? Surely that is not enough to make such a statement. Aristotle already noted this: anything can be said about birds, except that they are not endowed with language, or better, with *languages*, with different dialects and different skills in modulating them. So, language is not our specific characteristic: there is another conspicuous trait that distinguishes the human being, at least as opposed to birds—the hand.

Keywords Anthropology · Supplement · Recording · Documents · Baculum

M. Ferraris (✉)
Università degli Studi di Torino, Turin, Italy
e-mail: maurizio.ferraris@unito.it

S. Chiodo, V. Schiaffonati (eds.), *Italian Philosophy of Technology*, Philosophy of Engineering and Technology 35,
https://doi.org/10.1007/978-3-030-54522-2_2

2.1 The Hand

When it comes to the hand, as we know, Aristotle disagrees with Anaxagoras. For the latter, the human being is the most intelligent of the living because he has hands. The Aristotelian version is naiver and harder to believe: the human being was endowed with the hand *because* he is more intelligent. In other words, all animals are equal, but there are some that are "more equal" than others, because they have hands. Imagine the scene in which the Unmoved mover reviews all the animals and subjects them to intellectual and attitudinal tests, finally assigning the hand to the human being. The animals are lined up, each with its own characteristics and abilities, the human being showcases her/his sociability (expressed with handshakes and pats on the back), her/his heaven-sent language and her/his extreme intelligence. God examines them and questions them, and, pleased with the human's intelligence, rewards her/him with one hand, or rather two. Except the human would never have won the competition if she/he had not already been equipped with the hand, that is, with what was at stake in the first place.

There are therefore good reasons to think that this was not the case, and that Anaxagoras was indeed right. Aristotle's vision supposes some exact moment when humans received language—something like a Pentecost coming down from heaven, i.e. something extraneous to them—whereby they decided to create society to simplify their lives (which is questionable) and end the war of all against all (which is even more questionable). Instead, as Anaxagoras posited, the human being is the most intelligent of the living *because* he has hands, and therefore can implement practical manipulations that, with a process that lasted for millennia, eventually led to the development of intelligence. In this sense, Anaxagoras' definition of the human being would sound like this: the human being is the animal endowed with hands. Humanism would thus really be a "handism", a power to manipulate and deal with clubs, handles, missiles and of course also coins, manuals, Moleskine notebooks and mobile phones—which, with very telling Anglicism, are called "Handy" in Germany. As we shall see, only the hand could ensure the development of language, eventually determining the usefulness of mobile phones.

Yet Aristotle's scepticism about the civilizing functions of the hand still makes a good argument. Many anthropoid monkeys have hands as well as opposable thumbs. Indeed, they have two more hands than bipeds, which is why they are sometimes designated as "quadrumana". If Anaxagoras was right, the chimpanzee should be by far the most intelligent of animals. On the other hand, it would be a mistake to think, like Aristotle, that it was the increased size of the brain that determined the evolutionary advantage of humans, since we know very well that Cro Magnon had a bigger brain than Sapiens, and that the human brain is smaller than that of a dolphin and extraordinarily similar to that of a pig (which, again, explains many things).

Not the creation of syllogisms or norms, but the construction of technical devices is the first act of the naked ape, whose first concern is to eat, to be armed, to warm up (and therefore also to dress up). Knowledge and duty will appear long afterwards, as derived from technology. It has rightly been pointed out that the gap between the ape to the human being lies in the latter's technological capacity, which does not take place on the level of *homo sapiens*, i.e. of the rational animal, but on that of *homo habilis*, i.e. of the technological animal. Indeed, the human being's evolutionary success seemingly clashes with the very laws of evolution, because it decrees the prevalence of the least fit. So, it is not enough to focus on the hand to account for it: one must look at what the hand grasps. Despite having four hands, chimpanzees do not systematically use sticks. Of course, they can do so occasionally, but we have never seen an old chimpanzee leaning on a stick, and this is for the trivial reason that the species has not made a final choice in favour of the erect posture, and, if necessary, can go back to walking on four legs. What would it do with a stick? And what do we do with it? (Bell 1833; d'Errico and Backwell 2005; Derrida 1987; Gibson and Ingold 1993; Johanson and Edgar 1996; Napier 1971, 1980; Radman 2013; Tallis 2003).

2.2 The Stick

As children we are often told about the riddle of the sphinx. What animal has only one voice (i.e. one essence, which consists in being endowed with language) but walks in three different ways: in the morning on four legs, at noon on two, and in the evening on three? That animal, as we all know, is the human being, but the answer suggests that being endowed with language and *hands* is not enough: it is not the human essence, its fundamental *eidos*. The technical prosthesis (the stick) thus falls within the very definition of the human animal, something that does not apply to other animals (you will never find an old cat leaning on a stick). In Oedipus' riddle, in fact, it is not only about hands, or paws that eventually lose the function of grasping and are reduced to walking: it is about sticks.

Let's think about the scene through anyone of the many artistic reproductions available, from the Greek vases to Gustave Moreau's paintings. Usually, on the left there is the sphinx, sitting on a column or sensually wrapped around Oedipus. It is an evolved and hybrid animal, because it has the body of a lion and a human face. If the sphinx had kept its lion face, the scene would have been worthy of a comic book, because lions cannot speak and have a snout to carry out functions which, in humans, are carried out by their hands and technical devices. Now, in the treatise *On the Making of Man*, which dates back to the end of the fourth century, Gregory of Nyssa establishes an essential correlation between the acquisition of hands and the development of language (Corballis 2002), which Darwin would re-propose without variations fourteen centuries later (Darwin 2010). If humans did not have hands, then their faces, like those of quadrupeds (like lions), would have an elongated shape, lips suitable not for articulating words but for grazing, and a thick, calloused tongue good for kneading food (McGinn 2017).

Once again, the prerequisite for language is not so much a particularly developed brain mass (as would be logical following Aristotle and Heidegger's hypothesis) but rather the hand, in agreement with Anaxagoras and Derrida. The hand frees the mouth, the teeth and the tongue, making them available for speech. This shift should not be understood simply as a physiological development, but also as a technological, economic and social event. Indeed, unlike the mouth, the hand can hold a stick, and proceed to a series of capitalizations (as we will see shortly) that would be impossible if the mouth had to perform the functions of the hand.

To go back to the scene, let's now look at Oedipus. He is not old (he has dark hair and a beard in Greek paintings, in Moreau he has blond, long and wavy hair, something between Christ and a gigolo), but he always has a stick, in all the representations we have. Over time, this stick seems to grow longer than the singular stick of Pinocchio's nose, to the point of becoming a long red pole in Moreau's painting. The reason for this is simple: only the human being, who has chosen an upright posture at her/his own risk, really needs a stick. And since the upright posture has aroused dreams and needs completely different from those she/he had before, the human being has ended up using not only the stick, but an infinite number of technical devices (including the "leading-strings of judgment" mentioned by Kant) of which the stick is the fundamental emblem. Thus, the human being is the only animal whose definition necessarily involves a technical device. The human being is the ζῷον βακτηρίαν ἔχων, the animal endowed with a stick (and also infested more than other animals by those sticks we call bacteria), who is able to use the stick and the carrot with other animals, including other humans. Therefore it is the animal endowed with glasses, shoes, backpacks, mobile phones, books, the Internet, and receipts.

Claiming that the hand is an exclusively human element, Heidegger insists on the fact that, in addition to manipulating or handling things for practical purposes, hands fulfil symbolic purposes, too. This is true: hands are shaken, joined in greeting or prayer, stretched out in blessing or shaken in curse, kissed in court or in a mafia film. And kissing a ring, the bearer of the seal, is like kissing a signature. However, it is difficult to think that other animals with hands are unable to use them symbolically (what does the ape do when it beats its chest, in a gesture that unfortunately we also find in some politicians?). Above all, for our purposes, it is essential to note that this symbolic use also applies to sticks and tools in general. A urinal is an instrument, a work of art (if exhibited in a gallery) or a weapon (if thrown out the window during a riot). And for all we know, this symbolic duplicity may have started in the tool and only later been transferred onto the hand. Before the hand was used to make signs, humans and animals knew well that the trace of a paw indicates something that goes beyond simple presence: that is, it has a symbolic function.

Ultimately, the thing is very simple. The mere fact of having hands does not distinguish humans from other non-human animals, but rather indicates their specific shortcomings, precisely because humans have only two hands, and not four. The real difference is the stick: the stick of our old age, but also all the other sticks that fill our lives and the history of humanity as a whole. The human being has hands, sure, but the fact remains that without a stick she/he is an imbecile. She/he is an imbecile both in the technical and in the etymological sense, since the etymological root of imbecile is *in-baculum*, that is, without a stick ("stick-less"). So much so that for hundreds of thousands of years the human being, despite having hands, remained only a naked ape. The change took place when the hand finally grasped a stick. That is when the history of humanity and the history of technology began.

2.3 The Supplement

Quadrumana are much better than humans, in terms of hands, but—as noted—it is not simply a matter of hands, but of sticks. According to these premises, the Kantian question "what is man?" (which today we would reformulate as "what is the human being?") can be reworded as: what can be done with a stick, i.e. with the prosthetic definition of the human being? I propose to replace the canonical definition of the human being as an animal endowed with language with that of the human being as an animal endowed with a stick, i.e. with a technology of which writing and language are eminent but not exclusive manifestations.

The human animal needs supplements to make up for its deficiencies, and that is why humans have technology and technology has humans. A synthetic definition of technology could in fact be that of a "panoply of supplements", because technology is precisely the set of tools or weapons with which the human animal has equipped itself to make up for its shortcomings. These shortcomings—probably the most interesting aspect of the whole affair—would not have occurred if the human animal had remained in its place and accepted its biological destiny. In other words, what non-human animals can solve involuntarily and through very long evolutionary processes, the human animal solves through technology and society. Instead of an internal evolution, marked by genetic selection, the human being deals with an external evolution supported by technology and society.

The first function of the supplement is, of course, enhancement. The ape of *2001: A Space Odyssey* understands that it can increase the strength of its arm by using a stick (in this case, a bone). With that stick she/he becomes stronger than her/his peers. And the stick, which is thrown upwards as a sign of jubilation after the massacre and turns into a spaceship, shows the overall destiny of humanity and the path of technology. All the technical properties of the stick can be automated: strength (the most evident element of machines); authority (the machine imposes a procedure, the so-called "technical necessities"); knowledge (the machine increases knowledge through the archive, and knowledge has a mechanical element in itself; machine learning, which is based on recording, is the improvement of this approach thanks to automation); normativity (keeping track of the reciprocal relationships between social actors multiplies social objects and normative instances). Enhancement is only the visible part of a deeper situation of insufficiency.

Derrida commented on a famous photographic book on Heidegger (Meller Marcovicz 1978) that triggered a great invective by Bernhardt and focuses on Heidegger's reflection on the hand in an arc that goes from *Being and Time* to *What Is Called Thinking?*. On the basis of what I said about Oedipus and the necessary supplement, it should come as no surprise that these analyses, which obviously concentrated on the technical, philosophical and political role of the hand, require additional investigation. Specifically, one ought to place more emphasis on the supplement, the stick, which is all-pervasive and nearly more ubiquitous than the hand (Harman 2002). In the pictures, Heidegger takes the stick, raises the stick, rests the stick on a bench, hands an additional stick to his wife, uses the stick's knob to look as if he was sticking out his tongue, etc. And when the stick is at rest (but Heidegger leans on the stick even while sitting in an armchair and conversing with the journalists of the *Spiegel*), then the spotlight should go on the pencils and pens arranged in battle order on the desk. Indeed, in Heidegger's study in Freiburg, just as in universal history, it is the sticks, pens or barcode bars that have determined the most decisive revolutions. This is not because the spirit is everything, but only because the bars, as well as pens or that other form of sticks we call *bacilli*, manage to store a quantity of data that leave no trace on the stick used as a club, except some accidental notch left from a fight. Also, to leave a trace on the sceptre (another type of stick) requires the memory of the subjects, unlike the sticks holding the rolls of the Torah, which are the actual bearers of a memory that has been able to defy the millennia.

The stick is therefore what phenomenologists would call a reduction to essence. The stick is the necessary complement of the human, it is what makes up for the constitutive lack that differentiates humans from animals and makes them so ambitious and so dissatisfied. The stick takes us back to a much older past than we are willing to admit, more or less to the good savage imagined by Rousseau, but which (as we shall see) was perfectly aware of the need for give and take. That good savage, whether Rousseau liked it or not, was equipped with technology from the start. In fact, there are many superstitions surrounding technology: it is cold (which is paradoxical, if you consider that the use of fire is a form of technology), it is alienating (if only it were the case! Unfortunately it reveals human nature for what it is), we would be better off without it (which is obviously false: we would die at twenty, if we're lucky). But among all these superstitions, the most monumental and false one is that technology is modern, as old as plastic or at least stainless steel, as suggested by the expression "age of technology" with which both philosophers and non-philosophers like to define the contemporary world. Of course, this is not the case. Technology is as old as the human being, it was born when hominids began to transform into humans, which did not take place through an increase in brain mass (which is not necessarily successful from an evolutionary point of view), but through an exteriorization and enhancement of our skills through technology.

2.4 Externalization

There is nothing inside the human animal that justifies its difference from non-human animals. The difference is all outside: it lies in the hand that handles the tool, capitalizing strength (a stick fortifies the arm), memory (you can keep track of time on a stick, creating a calendar), and society (you can carve a notch for each transaction on two sticks, one of the debtor and one of the creditor, so that if you place them next to each other you can create an efficient form of accounting). In other words, the difference lies in the capital obtained through recording, which is transformed into an accumulation of culture, wealth and resources, whence the emergence of the purely human characteristic called project, born of an abundance of resources. Technology prolongs evolution and externalizes it, with a consequence of particular interest, namely that evolution no longer takes place in life, but in history; no longer in the brain, but in culture and society (Diamond 1978).

If this is true, it can only make us reflect on how inaccurate it is to speak of "instinct of language" and "collective intentionality". The first would allow us to speak, just as the bump of arithmetic allowed us to count (at least according to phrenologists). The second would be the foundation of the system of agreements and impositions of functions which gave rise to the social world. But, banally, if humans had not been able to light a fire, which requires certain skills with their hands, would they ever have been able to form the first communities? The latter developed from the acts of passing food (manipulating it), removing each other's fleas (with their hands), embracing and mating; in fact, they were still entirely similar to those of our animal past, the only difference was the ability to light a fire, which depended on the peculiar conformations of our hands.

Fire and the act of gathering around it would then cause other hand-related developments (Wilson 1998). We had weapons to strike, tools to weave and cultivate, and especially manufacturing practices that archived in themselves the processes that had formed them, giving rise to centuries-old traditions (think of the evolution of the artifacts of chipped stone). Whence the notion of history, which manifested itself primarily through representations (paintings, stones and statues that evoked past events), then developed into concepts, and finally became culture—a term that, however, let us not forget—draws its origin from cultivation, i.e. the technical manipulation of nature.

It is at this point that the human being had a language and a society. The latter comes from living together in a protected space marked by the sharing of goods, so much so that any failure to share and protect (politics gives us endless proof of this) generates a crisis in social solidarity, until the pact is broken and there is a declaration that society does not exist, but that there are only individuals. Language develops after the gesture, and after writing and manipulation, in the conditions made possible by both: it is a result and not a precondition. When the human being's hands are busy with technical devices, she/he will no longer be able to use them to express herself/himself, and she/he will begin to speak, in a process that lasts hundreds of thousands of years. Note that the earliest names in history indicated tools, just as the first surnames pointed to professions linked to manipulation: Miller, Taylor, Carpenter, not to mention the countless Ferrari, Smith, Schmidt, Lefèbvre, Herrera, Kovacs, and so forth.

2.5 Technodicy

One of the most interesting features of the supplement is the variety of forms it can take. The club was a supplement, but so are books, coins, garments. In all these cases, a technical accessory compensates for a lack in a non-accessory way. This lack, however, is not such from an ontological point of view, as theologians and psychoanalysts claim. The human being is neither ontologically defective nor ontologically prodigious and superabundant. It is simply a being who is not satisfied with its natural habitat, its lifespan, the possibilities that are written within its natural destiny. A being, therefore, who at a certain point decided to change its own evolutionary destiny through the use of supplements.

The human being is the misfit animal: she/he adapts to every environment precisely because no environment is apt to her/him, except for the African savannahs from which she/he originated, but in which she/he led a solitary, poor, nasty, brutish and short existence. And so in every environment she/he needs supplements, additions, integrations. Far from being the only owner of the world, the human being is the only animal radically poor in world, structurally unsuitable for any environment, except those places where humans, having no need for adaptation, do not even develop as humans. Thus, being an unstable, unfit and unsatisfied animal, only the human being is poor in world; precisely for this reason she/he needs technology, and above all those capital technologies that are writing, language and culture. Naturally defenceless, the human being develops late compared to other animals, and therefore more than these animals is in need of technology, parental care and social protection: we continue to develop even after birth, and we do not stop learning, and therefore being dependent, even when we have learned to walk and talk.

It is therefore ironic that the most dependent—and therefore least free—animal of creation should see freedom as one of its essential characteristics, which only increases its dependence and frustration. The human being is dependent for constitutive reasons, and the fact that freedom is conceived as one of its essential characteristics only increases its dependence. Freedom is a rare good, and is an ideal to be realized rather than a state in which we find ourselves (though in a defective form, affected by pathologies of the will). But every cloud has a silver lining: the mere fact of conceiving a missing limb determines an aspiration and generates an ideal, that of freedom not as a starting point, but as a point of arrival. In the relationship between the human and her/his supplements, a theodicy, or more precisely a technodicy, is thus outlined. It is precisely our inadequacies, and the supplements that they require, that determine the specific difference between humans and other animals: it is with the help of protection, enhancements and supplements that humanity becomes what it is. In other words, what we are depends much more on what is outside of us than on what is within us.

It follows that dependence on technology, as well as dependence in general, is a distinctive trait of the human species. Therefore, it is not a specific circumstance of our time ("the age of technology", the "power of technology" etc.), contrary to what is often suggested. Above all it is bizarre to think, as often happens, that humanity risks being eventually replaced by machines—as we shall see, this idea makes no sense because the notion of machine is only conceivable in relation to humanity. On the one hand, therefore, human subjectivity cannot be thought of without technical devices, as suggested by the theories of the extended mind, but as was already implicit in the classic descriptions of the link between identity and memory. On the other hand, in order to elaborate a theory of society, it is necessary to adopt a perspective that conceives it on a scale that is not simply human, that certainly takes into account our animal past, but that above all is capable of taking into account the technical and documedia system as an integral part of both social ontology and individual psychology (Epstein 2015). It seems banal, but when a mysterious and magical function such as collective intentionality is placed at the foundation of social ontology, this amounts to, rather paradoxically, talking about the social being regardless of technology and society (which survives only as an instinct and not as a historical reality).

2.6 Alienation

So far, so good. Let us pause for a moment to reflect on what we have said so far. It is very common to consider technology as alienation. According to this view, the essence of human nature, what we really are, is a synthesis of all virtues: we are fundamentally good, disinterested, dialogical, generous, and simple. Then we were corrupted through technology and society, which brought greed, lies, oppression, exploitation and many other misfortunes. This is the vision of humanity that comes from Rousseau and underlies the vast majority of discourses on technology: we are always on the phone, always arguing on social networks (in earlier days, we were always watching television), always texting instead of talking to our friends and family, what we have become, how we sank so low, etc. This amounts to saying that, if it were for us, we'd be completely different: evil comes from the outside in the guise of technology and transforms us, alienating us. Our original nature is perfect and autonomous, that is, we are the bearers of a Kantian morality where the legislated is also the legislator, and in any case is an autonomous source of law. In short, we are the bearers of Pentecostal meaning.

Obviously it is not so. It's true that we are always on the phone, but is it not because the human being is an animal endowed with language? It's true that we are always on social networks, but is it not because the human being is a social animal? Here's the point: technology is not alienation, but revelation—it shows humanity as it really is, beyond all self-deception, for better or worse. To hypothesize alienation means to think that there is a whole and complete humanity, which is then transformed and deformed by the intervention of something that comes from the outside. Alienation is a mythical and false idea, both when it refers to the fall of human nature from its state of perfection, and when it insists on the perversion introduced by science and technology (as Rousseau and many of his disciples did). Indeed, take the post-Fordist alienated worker, or project contractors who can go on holiday whenever they wish (except being summoned when they least expect it): are they really alienated? Of course we are not happy about constantly checking our phone for calls to arms but, at the same time, nothing in our lives reminds us of alienation in its classic meaning—fragmentation and repetition, and estrangement with respect to the products of one's labour. Indeed, what we see today seems to be the prototype of Marx's whole and emancipated human being, the human being who can "do one thing today and another tomorrow, to hunt in the morning, fish in the afternoon, rear cattle in the evening, criticise after dinner".

Also, note that "project" workers are indeed responsible for finishing a project they are in charge of themselves—quite the opposite of Charlie Chaplin, who had to screw the same screw over and over. If they are on holiday when they are confronted with a problem or some inconvenience that will ruin their weekend (or week off at the end of the project): the fact remains that this is a commitment to the totality of the project whose burden they took upon themselves; the mail or phone call therefore have some meaning within the organic project they are working on. In short, *mutatis mutandis*, the mobilized worker is not so different from a poet who finds sudden inspiration in the dead of night: like her, she/he might die young and poor, and like her she/he'll be the living proof that one can be non-alienated and still be unhappy.

2.7 Revelation

If this is the case, it makes no sense to speak of a humanity alienated by technology, since technology is a constitutive part of the essence of the human being. From the point of view of anthropology and social ontology, technology is therefore not a vehicle of alienation, but—as the ongoing documedia revolution shows—an instrument of revelation of the fundamental characteristics of human nature (of which technology, and not reason, constitutes the "second nature") and of the essential structures of social reality.

This consideration of alienation is loaded with consequences. One of them is that the human being, i.e. the only animal whose essence includes the technical supplement, is not alienated from her/his own essence because of the misdeeds of technology, as claimed by a die-hard (and, what is worse, false) naturalistic and primitivistic rhetoric. Human nature, instead, is revealed by technology: it is technology that tells us what we are and what we want. If you want to know what humankind is, look at its technology. This is so obvious and true that there are entire civilizations of which we have nothing left but technical devices (tools and furnishings), but it is enough for us to know something about them. Full knowledge begins when the human develops the ultimate technology of writing; more gloomily, what we produce through technology is a large amount of fossils, destined to survive after the disappearance of the human species.

Technology is therefore a paradox in the definition of human ontology. As such, any technological supplement appears to be an accessory to the human essence. And yet, since the human essence cannot be thought of without technology, the accessory becomes an ontologically constitutive element. This is not the only case of this additional ontology, on which Derrida has rightly focused his attention. Consider the status of the sign, which performs its function of reference, and therefore conforms to its essence, only if it partially renounces its natural essence. Ash is a sign of fire, but if we consider ash as a sign, then we add an additional meaning that in no way falls within the natural ontological definition of ash. But what, in the case of the sign, is simply a theoretical challenge, in anthropology is a fact full of consequences, with which it is much more difficult to deal. It is enough to think of all those anthropologies—the majority—which consider the human being independently of the system of supplements or, at best, only in relation to one supplement (society). Sometimes they even solve the problem by attributing to the human being an imaginary supplement (the soul, or collective intentionality) which has the incomparable advantage of being unanalysable.

Herein lies the role of revelation, and its content is precisely this: there is no "in itself" of human nature, there is no Pentecostal meaning, but there is a historical becoming where technology plays a constitutive role: we understand what we want and who we are by the technology we use. The revelation has not at all ended with the Apocalypse; it still goes on, precisely thanks to technology, which is the prosthesis of any construction by enhancing the natural endowments of humans, and constitutes a context of emergence as much as—and perhaps more than—society. What apparently takes us away from us (alienation) is what reveals us for what we actually are. The Ptolemaic vision, linked to Pentecostal meaning, follows the direction Humanity → Society → Technology. In this view, the definition of the human being (rational animal, social animal) leads to understanding the characters of the social world and the technological world where the first is conceived as a human construction, and the second as a human extension. Conversely, the Copernican revolution, which draws from emergent meaning, follows the direction Technology → Society → Humanity (Kittler 1999). If you want to know the soul of the human being, you have to start from the society she/he lives in; but if you want to understand the society she/he lives in, what gods she/he worships, you have to understand what technology she/he uses (which very often uses her/him instead).

However great the social changes brought by the Internet, its major effect lies in revealing the deep structures of social reality, which are much prior to the web's structures. Studying the Internet thus means carrying out an in vitro study of the social world—addressing humanity as it is and not as it claims to be—from an ideal, albeit opaque, position: a huge amount of data that is constantly updated, some of which are easily accessible thanks to search engines. The fact that these data are only partly accessible to users completes the analogy between the Internet and the social world with its characteristic non-transparency, and constitutes an additional challenge for analysis. That is why today we're in a better position than ever to the ancient injunction "know thyself" found in the temple of Apollo at Delphi. To know ourselves, to answer the question "What is the human being?" (anthropological question) you need to answer the question "What is technology?" (technological question). In agreement with an authentic Copernican revolution (the inverse of

Kant's, which was a Ptolemaic restoration placing the human being at the center of the universe), we avert our eyes from the I think and look at the World Wide Web (Bunz 2012).

2.8 Capitalization

The first and most fundamental function that can be performed by a stick is the capitalization of strength. This capitalization can be immediate, as when using a stick as a direct extension of the arm, or mediated, as when using a stick as part of a technical apparatus, be it a spear or a shovel. You can also use a stick to capitalize on the strength of another stick, like the spear throwers of which we have much archaeological and ethnographic evidence. Here the capitalization is direct, it does not involve a temporal shift, and is based on the physical principles of the lever. However, we must not overlook the fact that it is precisely this capitalization—which is both concentration and channelling—that underpins the enormous capitalization of strength and memory that is technology in general. The first machines capitalized on strength, the latest capitalize on memory. And between the two is no hiatus, but a profound continuity. In the beginning there was action, and in the action there was technology. Technology can enhance action in many ways: through the direct capitalization of strength, or through the capitalization of memory. The capitalization of memory is the origin of society, and limits the capitalization of strength by introducing another, much more powerful kind. And this does not stem from some contract or construction, but from the emergence of the essential forms of social reality, which are rooted in our animal past and are perfected and manifested in contact with technology.

After all, everything was already written in the ape of *2001: A Space Odyssey*: after testing the strength of his enhanced arm on the skulls of his mates, he throws his improvised club into the sky in jubilation, which is both brutal and social. As we know, the club seamlessly turns into a spaceship. Today it would turn into a smartphone. The history of technology and the history of humanity are inseparably linked, and their origin lies in that first act of capitalization. Not only wars, but architectures and industries as we have known them throughout human history originate from that first act of capitalization. The industrial revolution itself—in as much as it replaced human strength and hydraulic and wind energy with the power of steam, and then that of electricity—is fully part of the history of this capitalization. Of course, if one relied on a stick as the sole instrument to use in a fight, one would not go too far. Also, as the proverb goes, live by the stick, die by the stick. Humans soon realized this banal fact. Instead of being a simple tool for multiplying strength and acquiring authority, the stick can become the *symbol* of authority. Be it the direct and still effective threat of the policeman's baton, the king's sceptre or the conductor's baton, the stick is very well suited to the capitalization of authority, both as a deterrent and as a charismatic element. The chief, capital and the capital city of a country systematically need sticks, whether it be the king's sceptre, the bishop's pastoral staff, or the accountant's pen.

No matter how great the increase in strength conferred by a tool may be, it will never be as important as the general empowerment of the human species made possible through the joint action of a hand and a stick. The hand frees the mouth from manipulative needs, thereby handing it over to the production of meaning and expression. The stick, a technical device that was initially simple but became increasingly complex, created a safe condition in which the human being could develop meaning, social communication, and the exchange of goods. Just as authentic military power is not manifested in swords and plumes, but in the war room, so the greatest benefit and the greatest strength that is capitalized on by the stick does not lie in its direct action, but in the context that it produces. As a weapon or as a cultivation tool, as a prosthesis for walking or as a sign, the stick exerts an indirect action of incalculable scope: reducing immediate need and pressure, it frees the powers of intelligence and reflection. In this sense, the capitalization of strength constitutes the premise, negative but indispensable, of those other capitalizations—of authority, knowledge and duty—which characterise human civilization. The warrior exhibiting his strength in tournaments, the leader attending a parade, the priest blessing the emblems are not simply a social trinity deeply rooted in the Indo-European tradition. On the contrary, external deference to physical strength hides the awareness of the fact that brutal force constitutes the adverse foundation and the condition of possibility of intelligence.

Usually, when one thinks of the capitalization of strength, one only looks at this process from the point of view of enhanced energy. The hand and the arm become stronger if they can use a stick, a chipped stone, or a propeller. However, it is necessary to see the capitalization also from another point of view, that of liberated energy. By enhancing the hand, the instrument carries out the process of emancipation of the mouth and face that marks the beginning of the technological fortune of our species. If, like most other animals, our contact with the outside world took place mainly through our mouths, with our limbs responding essentially to the need for mobility, then our technical equipment would be radically reduced. Conversely, with the differentiation between hand and mouth, two different technological processes began and developed in parallel. From the point of view of the hand, it became possible to equip oneself with tools to enhance one's strength and resistance. From the point of view of the mouth, emancipation from manipulation needs translated into the possibility of using that organ in functions of symbolization and expression. There is therefore no contradiction in the fact that the human being is, at the same time, the animal endowed with hands and the animal endowed with language. However, it is important to note that the lack of hands would have determined the absence of language.

Therefore, the enhancement that comes from the stick and its derivatives or parallels, such as stones, is not reduced to the mere mechanical fact of the increase of strength based on mechanisms that are more or less those of the lever. More importantly, the hand shifts from being a tool (such is the case with bare hands) to being a motor or control system. This story culminates in the situation that Marx complained about, where the complexity of the machines eventually transformed the human hand into a mere supplement for the machines themselves.

2.9 Recording

An evolutionary circle is thus created. On the one hand, the hand relieves the mouth of onerous functions, and equips itself with technical tools of increasing complexity which make human life more suitable for the capitalization of goods, knowledge, and organization: hence the protection from both predators and bad weather increased social interaction, longer lifespan, and the development of symbolic capabilities and technical devices, such as writing, which allow humans to capitalize further. On the other hand, through language, the mouth enhances all the acquisitions of the hand: it facilitates social interaction by allowing people to speak while carrying out other technical tasks (for example, preparing food and chatting). Also, it confers an additional symbolic and social function on technological devices. In fact, a bivouac is much less interesting if it only serves for keeping warm, and it becomes much nicer if you add attractions such as fairy tales and songs around the fire, which in hundreds of thousands of years will lead to guitars, karaoke, barbecues and mojitos.

But this evolutionary circle, very easy to describe in retrospect, would have been impossible without that essential condition of capitalization that is the possibility of recording. Not one of the passages I have described would have been possible in a human universe incapable of keeping track. This ability, at first, is implicit and hidden behind more showy functions, such as the manual repetition of technical procedures, but soon manifests itself and takes over the scene. Notches, wedges, hieroglyphics, numbers and culture come to the fore, and humanity understands that the stakes are right there, in recording. For this same reason, the boom of writing in the documedia revolution is not an accident, but the revelation of something that was already taking shape when the virtuous cooperation—and separation—between the hand and the mouth began.

It may be interesting to note that different supplements all tell more or less the same story. The stick was born as a capitalization of strength but evolved as a capitalization of memory, with a process that spans over hundreds of thousands of years and is rooted in our animal past. It is worth noting, however, that the same process that starts from the strengthening of immediacy and comes to the capitalization of mediation also occurred in technical devices whose evolution took place under the eyes of my generation. Such is the case of the mobile phone, which was born as a talking machine, therefore as a prosthesis and megaphone of an immediate function, but very soon became equipped with devices of temporal deferral, becoming first a typewriter, then a recording machine. And so we've come to the current state of externalization of the archive, where the mobile phone is only the intermediary between the human user and the platforms responsible for recording document production. What one might wrongly interpret as a process of dematerialization or spiritualization can more correctly be understood as a process of deferral, the ultimate result of which is the strengthening of recording. In this way, the evolution of technology reveals to us the essence of civilization.

In general, it is problematic to conceive of any emergence outside a temporal dimension. This is what Derrida indicated with the term *différance*. The emergence of meaning also has a diachronic aspect. The vowel A, alone, does not mean anything. It only means something together with B and C, but this process of differentiation between arbitrary signs takes time, and therefore recording. One must imagine a three-way process in which recording allows for diachrony, and the latter allows for emergence. It is in recording that we find the three processes of inscription, iteration and alteration that allow for emergence. This specific importance of recording means that capital is only one of the many historical names assigned to the power of recording.

2.10 Emergence

"*It is now no longer the labourer that employs the means of production, but the means of production that employ the labourer*", lamented Marx (the emphasis is in the original). This statement suggests, as the rest of *Das Kapital* shows, that there should be a relationship between the human being and technology where the former is not alienated but in control. In truth, any kind of relationship with technology involves a form of submission, which therefore depends on the nature of the human being and technology, and not on capitalism. Now, anyone who makes a Nespresso must serve, and not only use, the machine, and this activity is not labour. Even artisanal work is conditioned by (and receives its purposes from) the tool being used, and this also applies to intellectual performance: confirming what has been said about "learning by doing", Euler said that his theorems were on the tip of his pen and Proust claimed he was not free in writing his books. Now, Marx cultivated the illusion of a free craftsman who was alienated by a cruel world. But if Proust and Euler were not free, then who is? While this is not the same as comparing Werther to an English miner (Goethe was well aware of the difference, and so are we), it seems that the relationship between the human being and technology must be interpreted, as I propose in these pages, as an essential form of revelation.

What is often an emergent tool, something that you find there and then to defend yourself, attack, or lean on, reacting to danger or fatigue, can be philosophically considered as an instrument *of* emergence, or as a form of preadaptation that ensures the emergence of essential traits of human nature and society. The stick as a universal supplement to humanity remedies our natural imbecility and allows for a series of acts of capitalization. First of all, as said, the capitalization of strength, because the stick can act as a tool, as a lever or as a club. Then, the capitalization of authority. There is no need to actually use the stick when we can suggest that we could use it at any time, and here the better stick is the club, the ancestor of the baton and the sceptre, rather than the lever, the ancestor of industrial and artisanal works. And since the pen wounds more than the sword, the club soon turns to other types of capitalization: the capitalization of memory in knowledge, the capitalization of acts in duty, the capitalization of traces in power. These are the most important capitalizations in a time like ours, in which recording reveals all its incredible, and long overlooked, power. And it is mainly these three forms of capitalization that underpin the pivotal transformation that constitutes the object of this book.

Technology, understood in the broadest sense as the transition from ontology to epistemology, can be conceived as a science of emergence. Emergence, as an evolutionary phenomenon, both in nature and in society, takes a very long time to unfold, showing an intrinsic correlation between emergence and temporality; however, the emergence of knowledge from being can be achieved much more quickly. Consider how quickly one can shift between forms of a bistable figure, or think of a detective's sudden intuition, the quick understanding that comes with experience, and now the data instantly processed by an algorithm. In this emergence, recording plays a key role. Here we touch on a metaphysical foundation that will be addressed in the conclusion of this book, but meanwhile it is essential to explain why recording is the necessary condition of emergence. Emerging means generating a quantitative accumulation of traces, the result of which is a qualitative change. For this to happen it is necessary that the traces are recorded, that is, that the accumulation takes place. And the condition of the process is obviously the possibility of recording. Recording is the most extensive transcendental, as it makes emergence possible. If we have recording, in fact, we can generate a system of relationships and differences that will lead to the emergence of meaning on the conceptual level, and of temporality on the level of intuition.

And once meaning has emerged, that's it. The hand, which is no longer a tool, but a motor and a driving force, generates a technical system that first takes care of finding energies other than human strength—so that the motor soon becomes the wind or water that moves the mill—and then, with a turning point that is the basis of the industrial revolution, the pressure of steam. Artificial intelligence is no exception, and is fully part of this transition from tool to motor to driving force. Also in this case, intelligence, which was originally a direct instrument of the human being, becomes a motor and a driving force because it provides artificial intelligence with its fuel, i.e. the database on which to run, while also determining the system of purposes and needs to which artificial intelligence is called to respond. Ultimately, however, the aim of the whole process consists in transforming the tool not only into a motor, but first of all into a driving force, that is, into the ultimate aim. This is the situation we are witnessing today. Freed from any productive necessity, the hand is essentially destined to writing, to controlling devices, to opening packages. Completely exempted from production, today the hand, like the human being, finds its *raison d'être* in consumerism.

References

Bell, C. (1833). *The hand: Its mechanism and vital endowments as evincing design*. London: William Pickering.
Bunz, M. (2012). *Die stille revolution: Wie Algorithmen Wissen, Arbeit, Öffentlichkeit und Politik verändern, ohne dabei viel Lärm zu Machen*. Berlin: Suhrkamp.
Corballis, M. (2002). *From hand to mouth: The origins of language*. Princeton: Princeton University Press.
d'Errico, F., & Backwell, L. (2005). *From tools to symbols: From early hominids to modern humans*. Johannesburg: Witwatersrand University Press.
Darwin, C. (2010). *The descent of man*. New York: Dover Publications.
Derrida, J. (1987). *La main de Heidegger (Geschlecht II), in Id., Psyché* (pp. 415–451). Paris: Galilée.
Diamond, J. (1978). The Tasmanians: The longest isolation, the simplest technology. *Nature, 273*, 185–186.
Epstein, B. (2015). *The ant trap*. Oxford: Oxford University Press.
Gibson, K. R., & Ingold, T. (Eds.). (1993). *Tools, language, and cognition in human evolution*. Cambridge: Cambridge University Press.
Harman, G. (2002). *Tool-being: Heidegger and the metaphysics of objects*. Chicago: Open Court.
Johanson, D. C., & Edgar, B. (1996). *From lucy to language*. New York: Simon and Schuster.
Kittler, F. (1999). *Gramophone, film, typewriter (1986)*. Stanford: Stanford University Press.
McGinn, C. (2017). *Prehension: The hand and the emergence of humanity*. Cambridge: MIT Press.
Meller Marcovicz, D. (1978). *Martin Heidegger: Photos: 23. September 1966-16. und 17. Juni 1968*. Stuttgart: Fey.
Napier, J. (1971). *The roots of mankind*. London: George Allen & Unwin.
Napier, J. (1980). *Hands*. London: George Allen & Unwin.
Radman, Z. (Ed.). (2013). *The hand, an organ of the mind*. Cambridge: MIT Press.
Tallis, R. (2003). *The hand: A philosophical inquiry into human being*. Edinburgh: Edinburgh University Press.
Wilson, F. R. (1998). *The hand: How its use shapes the brain, language, and human culture*. New York: Pantheon Books.

Chapter 3
Ethics and ICTs Beyond Analytic and Continental Philosophy

Adriano Fabris

Abstract This article aims to develop an ethical reflection on ICTs with an in-depth analysis on three topics that will introduce new elements to the history of thought. First of all, I will analyse the categorical difference between "technique" and "technology", and the novelties that this concept expresses. Then I will analyse the changes that occur today in the field of communication: in particular, the transformation of the function of communication, which has moved from data transmission to the construction of environments. Finally, I will outline the reasons why it is necessary today, in the face of these new scenarios, to rethink ethics as a philosophical discipline, by going even beyond the setting of the so-called "continental philosophy", as well as the approach taken by "analytic philosophy".

Keywords Ethics · ICT · Technology · Continental philosophy · Analytic philosophy

3.1 Three Fundamental Issues Regarding ICT Ethics

One of the fundamental issues that Information and Communication Technologies (ICTs) brings to light is the effects of these technologies on our behaviour and, more generally, on our lives. Another issue that arises directly from this concerns the need to identify the criteria and principles on the basis of which we can orient our behaviour and make our choices in a technological environment. These are two aspects that must be addressed by ethics and, in particular, by a specific applied ethics: the ethics of ICTs.

A. Fabris (✉)
Università di Pisa, Pisa, Italy
e-mail: adriano.fabris@unipi.it

S. Chiodo, V. Schiaffonati (eds.), *Italian Philosophy of Technology*, Philosophy of Engineering and Technology 35,
https://doi.org/10.1007/978-3-030-54522-2_3

In particular, it is a question of considering and evaluating, firstly, to what extent our behaviours change and, furthermore, how the world which we live in changes thanks to the development of these technologies. It is then a question of examining how we move in the new communicative environments, of identifying and justifying the shared criteria needed to act within them, and therefore motivating those who live in these environments to welcome these developments and make the right choices. The first form of study is referred to as the "ethics *of* ICTs" and the second as the "ethics *within* ICTs". They correspond to some extent to what in the context of contemporary ethics is called, respectively, meta-ethics and normative ethics: general expressions that are applied to the field of communication and technologies related to this area.

Within this ethical reflection on ICTs, it seems important to me that an in-depth analysis of this context should be made by taking into consideration three topics which, as I will try to show, constitute a real novelty regarding what human beings experienced during history. These are issues that concern, as follows: (1) the idea of technology in general; (2) the way we live communication relationships (not only data connections); and (3) the conception of "ethics" which, in relation to this, needs to be elaborated if we want to be ready to face the transformations of the present. As a result, over the past few years I have dedicated a lot of my work to these issues; more precisely to the fact that, in the technological age in which we live, the very concept of "technology" has taken on a very specific meaning and cannot at all be confused with the notion of "technique". In fact, I refer to the idea that today communication can no longer be understood only as a transmission of data from a sender to a receiver, that is as a process of information, but it has turned into a real environment in which we live and with which we interact. Finally, the fact that these conceptual and behavioural transformations require that the very concept of "ethics"—that is the discipline that reflects upon the criteria and principles of our actions, and that presents the problem of how to motivate the human being to act well—must be re-thought differently compared to the past.

There are, as we can see, issues that have great relevance from a philosophical point of view, since the conceptual, behavioural and mental transformations that ICTs have produced and continue to produce affect the ways in which the traditional notions of philosophy and even the very idea of being human have been developed until now. Today it is no longer possible to perceive this in the same way. Those notions and conceptions now appear outdated: both with the approach which takes into account the historical genesis of certain problems, typical of continental tradition, and that which follows the procedures and methods of investigation that characterize contemporary analytic philosophy.

It is therefore necessary to change strategy. We need to react to the new challenges brought on by technological developments by referring to the categories created in the past. Indeed, these very technologies—in our specific case the ICTs—are radically changing the behaviour of human beings and the very structure of the ecosystem in which we live. In short, we have to develop an adequate philosophy, one that can help human beings make the important choices that need to be made today.

3.2 The Difference Between "Technique" and "Technology"

In the following paragraphs I will analyse in detail the topics above mentioned. The first point for an in-depth analysis regards the distinction between technique and technology, which are necessary for a better understanding of the world we live in and how to interact with it correctly. These are two different forms of action, linked in different ways to how people work. Technique is what widens the possibilities of human action and which depends on human beings for its use and the development of the tools that put it into operation (Gehlen 1988). Technology is also functional to this extension of human possibilities, mostly carried out on human input. However, it is important to consider the fact that within the technological environment the devices used are acquiring an increasingly stronger ability for self-regulation—a sort of "autonomy", that is the possibility of modifying their outputs depending on the environmental inputs they receive—so that they are no longer completely under our direct control (Dusek 2008). This is the real leap in quality, the change of paradigm on both a cognitive and a practical level, which elapses between the sphere of technical instruments and that of technological devices.

Let us analyse this distinction better. The use of the term technique implies the assumption, by the human being, of his/her own limits. Somehow we need to be aware of these limits if we want to find a solution. This solution, however, does not necessarily resort to specific qualities of the human being, or to natural qualities that one may acquire, but rather to something that is external to the human being and that he or she can use. To reach the leaves of the highest branches a giraffe relies on its long neck, whereas a person uses a stick to collect the fruits at the top of a tree. As a result, the entire world is transformed into a reservoir of possible tools that serve to compensate for the lack of possibility.

In order for the object used to become an instrument, however, it must be subordinated to a purpose. It must be understood as a means to an end. And it must be functional for this very purpose.

It is the human being who sees the purpose, who conceives the object as a means to it and considers its congruity with the purpose that he or she intends to achieve precisely through the use of this medium. The human being, in other words, transforms the very concept of that object. Something as simple as a stick on the ground can turn into an instrument: a stick as an extension of an arm. In this way the human being gives this object a meaning that subordinates it to the perspective that the human being is able to indicate and for which the simple thing, which has become a means to achieve a purpose, can serve.

However, throughout history, and particularly in that of science and technology, there is a progressive change in the status of the technical tool and, consequently, of the relationship that the human being has towards it. The various technical tools, in order to better achieve their goals and to achieve increasingly more complex goals, and in order to progressively extend human capabilities, are coordinated among each other and sorted into a system. We are therefore witnessing an increasing quantity of complexification, which concerns instruments that up to a certain point were isolated from each other (Singer 1957). Now, these tools can perform more and more elaborate functions by connecting them in a system. Think of the cogwheels, the springs, the levers that, connected in a unitary mechanism, allow a watch to tell the time.

From this process, which develops in ever more sophisticated forms, an unexpected consequence has derived, a sort of "heterogenesis of the ends". In fact, the more complex and articulated is the system of technical tools, connected to each other in order to achieve a specific purpose, the more their functioning and the achievement of this purpose are beyond the control of the human being. And the more the human being is assisted by complex systems, the more useless his/her intervention and his/her initiative are.

Consequently, the technical tool tends to refer more and more to itself and in this way it ends up excluding the human being from its own processes. This has caused a new phenomenon by which the technical means are not only subordinated to other technical means, within a technological system, but the ultimate goal of their being connected in a system goes beyond what they were originally programmed for by the human being.

The result is a surplus of power that also goes beyond an increase in that power and performance obtained for different instruments put together and coordinated in a system. This surplus of power is linked to that "self-reference", and that "autonomy" that is acquired by the system of the instruments. Pay attention, again: here the term "autonomy" does not mean "human autonomy", that is the kind of freedom which belongs to human beings, but it refers to the structure of self-learning systems, that are able to modify their programmed activity depending on the environmental inputs they receive. In this new scenario however human beings are no longer really dealing with simple tools anymore. These instruments connected in a system, where they are able to refer to themselves and, within certain limits, to self-regulate and interact with their environment, are properly transformed into technological devices. Take for example the Google car.

The word "technology", in short, indicates a set of phenomena that is substantially different from that produced by technical instruments. Technological devices are something that, despite being built by humans, they end up gradually emancipating themselves from us. The driverless car or the robot capable of adapting its performance to what it relates to—that is, to "learn"—are examples of devices that, despite being built, programmed, repaired by the human being, perform their activity independently. That is, they are able to act without receiving constant human input.

There is therefore a real paradigm shift, as mentioned above, between the technical tool and the technological device. And this paradigm shift also involves a radical change in the relationship that develops between human beings and the world around them. This relationship is no longer that for which, for example against unexpected natural events, the human being can use technical tools. Now not only is nature becoming increasingly more artificial, but it is an area that is integrated, and often replaced, by technology produced by us, but which is not always kept under control. The setback caused by natural events is accompanied, and in many cases replaced, by the setback caused by the increasingly "autonomous" action of technological devices, in the abovementioned meaning. As a result, human beings need to come to terms with all of this and to adapt to the changes.

3.3 ICTs as Environments

However, there is a second new element, which characterizes the world in which we live and which calls for a philosophical reflection that can live up to the changes taking place. It concerns a particular consequence of the technological processes, considered in their substantial difference with respect to the form of action through technical tools. The technologies, in fact, in the excess of power that they generate and in the self-referentiality that characterizes them, are able to give shape to specific artificial contexts of relationship. In other words, they are able not only to influence the environment in which we live in, but also to create new environments in which we can insert ourselves and with which we have no choice but to interact. Let us consider the automated mechanisms of economic transactions, in which most of the processes and decisions are carried out without human intervention, even though they have serious repercussions on human beings. Let us consider not only the transformation, due to the technological devices in healthcare environments, but above all the creation of that actual limbo between life and death in which specific machines can make people survive even in terminal illness. Let us think, more positively, about the many real-life and virtual environments that are created by ICTs (Wahlsham 2001).

In fact, I would like to concentrate, above all, on what is brought about by these new technologies. There are many aspects that need further analysis. The first concerns the transformation in this scenario of the very concept of "communication" and of the function that communication itself assumes. Communication, in fact, is no longer primarily and simply a space of sharing, nor is it simply resolved in a transmission of data or information. Communication constitutes rather an environment in constant development: in other words, it creates the very environment which we live in.

Communicating today does not only mean transmitting data or news, but living in an environment. More precisely, it means interacting with data and information transmissions that, as a result of their activity, produce a specific reality. This is evident considering the fact that today technological devices—for example computers, tablets, smartphones—constitute the gateway to areas that we now live in every day. These are areas that are constantly fed by the passage of new information.

This refers to the "infosphere", understood as the global sphere of information in which we are immersed (Floridi 2016). Scholars, in this regard, underline the enormous amount of data that we can intercept and the global state of their diffusion: that is to say their constant expansion in time and space. However, this is not the only aspect of the issue. In fact, in addition to the environments created by the transmission of information that concern or expand a reality that we can still experience on a daily basis, there are other environments, also made possible by the programming of specific communication processes, which imitate everyday reality, extend its boundaries and at some point come to an end, in many cases, to replace one with the other. This refers to virtual worlds.

It is not possible to dwell here on the specific characteristics of the so-called "virtual reality": an expression that would sound like an oxymoron, if we were not used to it by now, considering this dimension itself as effective, that is, as something "real". This has already been mentioned (Fabris 2018a, chapter 3) and explained how it can be fully understood only by putting and linking together traditional concepts such as those of "possibilities", "potentiality" and "power" (in the sense of "strength"). They cannot then be opposed to those related to "reality", "actuality" and "passivity", to the extent that the power of virtual reality is precisely that of overcoming such contrasts and of making, so to speak, a synthesis.

I believe there are two important things that it is worth stressing at this point and which reaffirm the novelty of the processes in progress and the need we have to consider them as philosophical categories sufficient enough to uphold these novelties. Indeed, (1) it is true that in the environments set up by information processes the ancient idea of communicating as creation and sharing of a common space, in which the various interlocutors find the possibility of interacting with each other and with space itself, is re-proposed. This is the idea that refers to the etymology of the word "communication"—which refers to the Latin *communis*: common, belonging to everyone—and which makes it possible, among other things, to understand the rhetorical developments of the act of communication (Manetti and Fabris 2011). And yet what happens today is something more and very different, since the communicative space, real or virtual, is that opened up by technologies.

Consequently, (2) for the reasons analysed in the previous paragraph, we are no longer the ones controlling these processes, but we find ourselves fully immersed in them. If in the past the virtual reality disclosed, for example, by a novel was the result of the encounter between the imagination of the writer and the imaginative capacity of the reader, and if therefore it was enough to close the book to fall back into everyday reality, today the flows of communication are very different due to their quantity and their specific characteristics, and that the need for such distinction and the need for verification itself are abolished. In fact, not only does online reality tend to take the place of offline reality, but also this is now so invaded by information and stories—the so-called "storytelling"—that is very difficult to keep them under control. The proliferation of fake news is a result of this situation.

However, there is an even more important aspect regarding the transformation of communication in the age of ICTs. Today, I repeat, it has become clear that it is no longer possible to reduce communication to a simple transmission of data from a sender to a receiver (Shannon and Weaver 1949). Communication, even as a mere transmission, still foresees a surplus. This surplus has a symbolic character: it exceeds the pure and simple act of sending and receiving information.

It is thanks to this aspect that communication becomes an environment. In fact, what it entails is the opening, the development, the maintenance of a context of relationships. These are possible relationships that the ICTs put into practice and that make them accessible to human beings. Of course: it is true that what is achieved by ICTs is due to the programming activity and the insertion of content by human beings. At the same time, it is necessary that there are human beings who know how to understand this symbolic character of communicating, who know how to take advantage of this inherent surplus in every transmitted information and who can interact with it, as users, or who can create these spaces of meaning, for example as programmers of a website or as users of a social network. And yet the results achieved, right in front of our eyes, would not be possible if there was no intervention to concretize them and to keep the power of ICTs going on. It is a power that transcends the ability to enter data or transmit information. It is for this reason that human beings, who are not in control of these technological processes, find themselves *interacting* with these environments, and not just *acting* on them.

This represents a new point that needs to be emphasized. It is true that throughout the course of history human beings have always found themselves constantly interacting with natural events, and indeed have become used to the idea of having to face the unpredictable consequences of these events. Technique, as we have seen, served precisely for this purpose. The development of human culture has been realized in this direction, transforming the natural world into something more artificial and, indeed, replacing nature with artificial structures in order to place it under human control. But not only has this purpose never been fully achieved, as recurrent ecological catastrophes show. Above all, it has been proved that the way in which we tried to acquire this control, through technological developments, caused new uncertainties and new insecurity. Human beings therefore find themselves today faced with both the unpredictability of natural phenomena, only partially the object of a possible prediction, and the growing autonomy of environments created by communication processes. That is why action represents more and more a form of interaction.

There is one last aspect, then, which must be emphasized and which still concerns the transformation of communication in a living environment. In fact, this environment is never just one. This emerged when, in my previous explanation, I often used the plural, speaking precisely of "environments".

In fact, today we are citizens of multiple worlds and we can move from one communicative environment to another, thanks to what technologies offer us, and not only. We can move from an interaction with an offline reality, and with what this reality offers us, the chance to follow one or more online realities, in which and from which to navigate. We are truly citizens of multiple worlds: another characteristic that, in this form, has never been experienced before now.

We must therefore learn to move in these multiple environments. It is not just a question of knowing how to live within them, but also, and above all, putting one in relation with the other: a relationship that is valid and good for us. In other words, what are the choices that we can make *within* these environments and *among* these environments? It is a choice that requires the elaboration of very precise criteria and principles so that it can be adequately accomplished. This is also why there is the need to develop the ethics of ICTs.

3.4 Ethics and ICTs

Most of my recent research has been devoted to developing this specific kind of ethics. As mentioned at the beginning, I started from the idea that today it is necessary to define the criteria and principles on the basis of which we can orient our behaviour and make our choices within a technological environment. This implies, above all, the need to distinguish ethics *of* ICTs from ethics *within* ICTs. The first form of reflection, understood in its most precise meaning, has the task of studying and justifying the attitudes and customs that develop in the relationships we have with the communicative environment created by technologies. The second form has instead the aim to guide us in our concrete actions, or in the interactions that we have within technological environments.

More precisely, the ethics *of* ICTs aims to study the various forms of action that are inherent to information and communication technologies, the ways in which this action transforms the world in which we live and affects the behaviour of human beings. In particular, it not only analyses the processes of ICTs, but also studies the criteria to which these processes explicitly or implicitly conform and the consequences to which they lead surfaces. Think of the digital divide, the cultural divide, and all the transformations in mentality and access to knowledge that technological processes entail. Specifically, as can be seen, this form of ethics analyses the dynamics of ICTs, highlighting their principles and aims—whether they are attributable to human choices or, to put it better, embedded in technological mechanisms (Flanagan et al. 2008)—evaluating the possibilities we have as human beings to interact with them.

Ethics *within* ICTs, for its part, regards human behaviour inside of technological contexts, the possibilities and limits in interacting with these contexts, and the criteria that can guide us in the new environments we live in. While the ethics of ICTs, as I have already mentioned, bears some resemblance to meta-ethics (even if now, in reality, it is a matter of analysing not only the language of morality, but also, more generally, the ways in which we reflect on complex ethical situations, animated by the actions of artificial entities) ethics within ICTs more precisely concerns normative issues, i.e. it seeks to answer the question of how one should behave in those situations. These answers can be developed at various levels, from that of a simple etiquette, for example the so-called *netiquette*, to that of professional ethics—linked to the set of duties that can be formulated to regulate the activities of various ICT operators—up to and including the development of criteria and principles able to guide anyone who has to deal with technologies (Chiles 2013).

There are two new elements regarding this point. The first consists of an effective enlargement of the scope of application of the term ethics. We have seen that not only human beings, but also artificial agents act in technological environments, such as ICTs. We have considered how, as a result of the growing ability of these artificial agents to interact with their environment and to regulate themselves, their actions have become increasingly "autonomous", in the abovementioned meaning, with respect to the simple execution of what they were built and planned for. Consequently, the problem of an ethics concerning the action of artificial devices must be explicitly posed: in our case, those involved in information and communication processes.

This is the first new element, which emerges from an accurate analysis of the ethics of ICTs. The latter, in fact, cannot be simply understood as an extension of human ethics, to the extent that, as we have seen, the technological device is in turn capable of acting with a certain autonomy. The problem is at what levels and in what forms this autonomy can be subjected to an ethical scrutiny.

A first level, for example, is that of an action performed by a technological device that it has been programmed to do so in a certain way: even without being under the direct control of a human being. In this case the manufacturer and the programmer are still considered responsible for the action carried out by the machine. A second level is that of an action performed by a technological device able to elaborate decision-making processes through an implementation of its abilities, i.e. after interacting with its environment and "learning" from this interaction. In this case, in the operation of the device itself, it is also possible to identify a certain degree of "responsibility" and to attribute this "responsibility", in a broad sense, to artificial agents (Wallach and Allen 2009). What is not possible, however, is to believe that the technological device can consider and evaluate from an ethical point of view the activity it performs: that is, that it can be conceived, and that in turn it conceives itself, as a moral subject. What is missing is the possibility that this device refers to the criteria and principles of its own action, that is, that it critically takes feedback from them and that it is able to modify them, or make other choices. In other words, at present, a technological device performs and is able to implement the activities for which it was programmed, but not to self-program itself.

The second new element that emerges here concerns no longer the action of the technological devices and the ethical limits that characterize them, but the transformation of human ethics itself, once it is placed in relation with the actions performed by technological devices. I have already said that in these contexts the action of human beings increasingly assumes the character of interaction, rather than that of an action enhanced by the use of an instrument, as was the case in a technical context. Here again, a growing impotence of human action emerges, which manifests itself more and more as the power of technology unfolds. Above all, however, the centrality of the idea of a relationship is profiled, in the sense of the relationship between human action and technological action, as a topic to be explored and oriented in an ethical way.

From this point of view, the traditional structures of ethics show all their limitations, precisely because they are aimed at guiding and orienting a simple human act. In this context, then, it is useless to make reference to something as an essence or a "nature" that refers to beings like us, since in the technological field a significant role is played instead by the action of artificial entities. Neither the consequentialist approach can offer us effective help, since—as I mentioned—the consequences of human action, already difficult to foresee, become even more unpredictable within complex artificial environments where both human and artificial agents interact. Finally, not even a deontological principle can be useful, because once again it refers to the motivated assumption of a duty: this is the prerogative of human action and not that of devices.

If it is a question of *assuming* responsibility for certain actions—both in the sense of responding to the consequences of these actions and in the sense of responding to the criteria and principles that can pre-emptively guide them (Raffoul 2010), then this proper notion of "responsibility", I repeat, can be referred only to human beings. Only human beings can be aware of the consequences of their actions and of the principles that guide them. Only human beings can consciously refer to that principles and choose them. Both prerogatives belong to us. To say it better: they are characteristics that most clearly distinguish human beings from technological devices.

3.5 Beyond the Distinction Between Analytic and Continental Philosophy

In these new scenarios also certain distinctions, by now taken for granted in philosophy, are put into doubt. I refer especially to the distinction, predominant in the contemporary reflection, between "continental" ethics and "analytic" ethics, as aspects of a philosophy equally divided, in turn, into "continental" and "analytic" (Prado 2003, Chase and Reynolds 2013). Furthermore, what needs to be rethought, within the new scenarios that I have illustrated so far, is not only the meaning of the distinction, but also its presuppositions. Along with this the risks inherent in both the "continental" approach and the "analytic" approach must be highlighted. This has been the subject of reflection for some time (Bell et al. 2016). But even more it is necessary to discuss this distinction in the light of the ethical consideration of the ICTs as I have tried to develop so far.

First things first, let us consider the distinction between the two styles of thought defined respectively as "continental" and "analytic" as already something strange in itself, if one takes into account the fact that these two adjectives refer to, firstly, a geographical location, and secondly, to a methodological approach. It is therefore a distinction that does not have the character of a comparison, since the terms of this comparison are not placed at the same level.

However, the distinction exists. The continental approach is that which, when dealing with a philosophical problem, it refers to the history of thought, it interprets it, it takes it up again and re-elaborates it, always keeping in mind its history and referring back to it (Critchley 2001). The risk that the continental approach runs is that of remaining too bounded to the past and its tradition, and at the same time it limits the creativity of thought on the simple interpretative activity: an activity that, moreover, is often considered something that always depends more on the perspective and individual preferences of the interpreter. The analytic approach, for its part, is that which, in order to face certain issues, relies instead on an analysis of the problems conducted with logical instruments, on argumentative procedures that can be shared (rather than on identifying alternative points of view able to illuminate reality in different ways from the usual), and conveys the need for verification that is sometimes too concrete, made up of examples and counter-examples. The risk that this approach runs is not only to subordinate the experience of the past interested mainly in the present, thus accepting uncritically what is conveyed by the common mentality, but above all that of considering in many cases the whole activity of thought—and, in the case we are examining, the execution of moral action—as a process, indeed as a procedure, the validity of which depends on the ability to correctly follow certain rules (see for example Birsch 2014: pp. xv–xvi).

If what I have just said is true, then the scenarios opened by the reflection on ICTs and by the awareness of the ethical consequences, that their action entails, offer us the possibility to go beyond the limits of both the so-called continental ethics and the analytic one. We have seen the fact that we are now moving into a technological dimension, that is to say that today artificial entities are also taking action. We have also outlined the fact that, especially with regard to ICTs, we find ourselves living in an environment, indeed in a multiplicity of real and virtual environments. In the new context in which we live, therefore, ethics requires to be set up in a different way than in the past: ethics has to be related not only to human behaviour, but above all to the interaction, to the relationship between human action and device action. This ethics makes it possible to rethink the forms of ethics developed in the history of thought: the ethics developed by referring to human "nature", or to a consequentialist perspective, or to a deontological approach. From this point of view, turning to tradition, as does continental ethics, cannot offer solutions that can meet the needs of our time.

On the other hand, the task of this discipline cannot only be identifying specific criteria of behaviour, analysing the concepts that express them, justifying them by starting from concrete examples and indicating specific procedures for their application: as for example the various analytic theories of action seem to do (O'Brien 2015). This cannot occur because ethics cannot be resolved in the elaboration of well-defined processes and rules which need to be followed in a proper way. If we follow this path, in fact, we could end up disclaiming what it has characterized in the past and which still characterizes the exercise of an investigation on the criteria and principles of action, and on the motivations that drive us to act well. I refer to the creative capacity that is required by the application of these principles to concrete situations, and which today is increasingly necessary and evident in the broad and articulated sphere of the so-called "applied ethics".

The limit of the settings of analytic ethics, and its inadequacy in relation to the consequences of technological developments, are therefore very different from that which would be specific to continental ethics. While in the latter case the limit would consist in neglecting the specificity of the action of devices, since these ethics take into consideration only the tradition of human action, in the former case—which manifests itself in many developments in analytic ethics, above all of a normative approach—it would appear to be the attempt to bring human action back to well-defined rules and procedures. This would entail the risk, for this action, of being conceived in the same way as a mechanical process would, and even to a further extent. It would lead to the conviction that this type of process can represent not only the cognitive structure, but also the model to follow both for understanding and for the fulfilment of human action.

Today the idea that human actions have to be explained and guided in accordance of previously defined rules and procedures has a great appeal. The results reached by the technological devices, for example in the field of ICTs, certainly induce us to consider this procedural model as the most effective for understanding and justifying actions—also human actions. The result is a series of consequences. First of all, the possibility to treat both human action and technological devices in a uniform and unitary manner is obtained. Thus a gain is realized in terms of efficiency in the elaboration of the theory and the conviction is consolidated that it is no longer necessary to refer to the problem within the specific relationship between the two forms of action. Finally, it emerges implicitly or explicitly the idea that the only criterion that governs an action is that based on the effectiveness and efficiency of a procedure: therefore, a criterion that can be traced back to the principle of utility.

As a consequence of this idea, it is also thought possible to resolve the issue of motivation to act well, rather than badly. It can be done, for example, by offering an explanation of the processes of choice that allow us to reconstruct them and reproduce them. The ultimate consequence of this series of reflections is the reduction in moral behaviour towards mechanisms that can be studied scientifically—for example by neuroscience—and, ultimately, the affirmation of the futility of an autonomous ethical research. At the most, such mechanisms can be regulated by rules established and legitimized within a legal context. The ethical dilemmas therefore end up being resolved by laws and laws to be decided by parliamentary majority.

Therefore, compared with technological developments, continental ethics ends up being outdated and analytical ethics seems to lead, in some aspects, to a dissolution of the same autonomy of this discipline. From this point of view, although in different ways, the two approaches seem to indicate the fact that ethics has come to its end. On the other hand, however, the consequences of technological developments on the various spheres of life of the human being (that which bioethics and public ethics deal with even if in a different way) and on the whole ecosystem and its global processes in which the human being is part of (including environmental ethics and the ethics of economics but also the ethics of ICTs) need to be managed from an ethical viewpoint. Applied ethics, as previously mentioned, tries to give an answer to this necessity (Fabris 2018b).

To do this we need however to go beyond the rigid approach of both analytic and continental ethics. We need to start by rethinking the exercise of "application" in the field of applied ethics. By doing this, we realize that this application is not resolved exclusively in a mechanical procedure and that human action is not identified by simply following some rules. And if this is the case, it becomes necessary to distinguish again between the action of the human beings and the action of the devices. It becomes even more necessary to think how the relationship between these two forms of action can be performed in the best way.

This is the task of ethics in the technological age. In this perspective it becomes necessary then, as I have already said, to develop a *philosophy of the relations* within which it is necessary to insert the ethics of ICTs. It is not only a question of developing ethics of human relations in general, but above all, and more specifically, it is necessary to deepen the interactions between human action and the action of artificial entities, even in their differences, and to define the criteria that make this interaction good and morally justified.

I have already started doing this in some of the books I have written in Italian and in English (see especially Fabris 2016 and Fabris 2018a). In this essay, I have elaborated this project with a specific reference to the case of ICTs, and certainly much more needs to be done. Therefore, my following research projects will be dedicated to this issue.

References

Bell, J. A., Cutrofello, A., & Livingston, P. M. (Eds.). (2016). *Beyond the analytic-continental divide. Pluralist philosophy in the twenty-first century*. New York: Routledge.

Birsch, D. (2014). *Introduction to ethical theories. A procedural approach*. Long Grove: Waveland Press.

Chase, J., & Reynolds, J. (2013). *Analytic versus continental. Arguments on the methods and value of philosophy*. New York: Routledge.

Chiles, D. (2013). *Principles of netiquette*. Scotts Valley: CreateSpace Indipendent Publishing Platform.

Critchley, S. (2001). *Continental philosophy. A very short introduction*. Oxford: Oxford University Press.

Dusek, V. (2008). *Philosophy of technology. An introduction*. Malden: Wiley-Blackwell.

Fabris, A. (2016). *RelAzione. Una filosofia performativa*. Brescia: Morcelliana.

Fabris, A. (2018a). *Ethics of information and communication technologies*. Berlin: Springer.

Fabris, A. (2018b). *Etiche applicate. Una guida*. Roma: Carocci.

Flanagan, M., Howe, D. C., & Nissenbaum, H. (2008). Embodying values in technology. Theory and practice. In J. Van den Hoven & J. Weckert (Eds.). (2008). *Information Technology and Moral Philosophy*. Cambridge: Cambridge University Press.

Floridi, L. (2016). *The fourth revolution. How the infosphere is reshaping human reality*. Oxford: Oxford University Press.

Gehlen, A. (1988). *Man. His nature and place in the world*. New York: Columbia University Press.

Manetti, G., & Fabris, A. (2011). *Comunicazione*. Brescia: La Scuola.

O'Brien, L. (2015). *Philosophy of action*. New York: Palgrave McMillan.

Prado, C. G. (Ed.). (2003). *Comparing analytic and continental philosophy*. Amherst: Humanity Books.

Raffoul, F. (2010). *The origins of responsibility*. Bloomington: Indiana University Press.

Shannon, C., & Weaver, W. (1949). *The mathematic theory of communication*. Urbana: Illinois University Press.

Singer, C. (Ed.). (1957). *A history of technology*. Oxford: Clarendon Press.

Wahlsham, G. (2001). *Making a world of difference. IT in a global context*. Chichester: Wiley.

Wallach, W., & Allen, C. (2009). *Moral machines. Teaching robots right from wrong*. Oxford: Oxford University Press.

Chapter 4
The Ontological Interpretation of Informational Privacy

Luciano Floridi

Abstract The paper outlines a new interpretation of informational privacy and of its moral value. The main theses defended are: (a) informational privacy is a function of the ontological friction in the infosphere, that is, of the forces that oppose the information flow within the space of information; (b) digital ICTs (information and communication technologies) affect the ontological friction by changing the nature of the infosphere (re-ontologization); (c) digital ICTs can therefore both decrease and protect informational privacy but, most importantly, they can also alter its nature and hence our understanding and appreciation of it; (d) a change in our ontological perspective, brought about by digital ICTs, suggests considering each person as being constituted by his or her information and hence regarding a breach of one's informational privacy as a form of aggression towards one's personal identity.

Keywords Digital ethics · Informational privacy · Infosphere · Ontological friction · Personal identity

4.1 Introduction

"One of these days d'you think you'll be able to see things at the end of the telephone?' Peggy said, getting up." She will not return to her wondering again, in the remaining pages of Virginia Woolf 's The Years. The novel was published in 1937. Only a year earlier, the BBC had launched the world's first public television service

This chapter was originally published in *Ethics and Information Technology* (2005) 7:185–200.

L. Floridi (✉)
Oxford University, Oxford, UK
e-mail: luciano.floridi@oii.ox.ac.uk

S. Chiodo, V. Schiaffonati (eds.), *Italian Philosophy of Technology*, Philosophy of Engineering and Technology 35,
https://doi.org/10.1007/978-3-030-54522-2_4

in London, and Alan Turing had published his groundbreaking work on Turing Machines (Turing 1936).

Distracted by a technology that invites practical usage more readily than critical reflection, Peggy only half-perceives that new ICTs (information and communication technologies) are transforming society profoundly and irrevocably. The 30s were laying the foundations of the information society. It was difficult to make complete sense of such a significant change in human history, at this early stage of its development. Nevertheless, an evocative phrase concerning the topic of this article appears in an essay on Montaigne, again by Virginia Woolf (The Common Reader, 1925): "[we], who have a private life and hold it infinitely the dearest of our possessions [...]", will find protecting it ever more difficult in a social environment increasingly dependent on Peggy's futuristic technology.

Today, the commodification of ICTs, begun in the 70s, and the consequent spread of a global information society since the 80s, are progressively challenging the right to informational privacy, at least as westerners still conceived it in Virginia Woolf's times. The problem is pressing.[1] It has prompted a stream of scholarly and scientific investigations, witness this special issue of Ethics and Information Technology; and there has been no shortage of political decisions and legally enforceable measures to tackle it.[2] The goal of this paper, however, is not to review the very extensive body of literature dedicated to informational privacy and its legal protection, even in the relatively limited area of computer ethics studies. Rather, it is to argue in favour of a new ontological interpretation of informational privacy and of its moral value, on the basis of the conceptual frame provided by Information Ethics (Floridi 1999, 2008).

4.2 Informational Privacy and Computer Ethics

Why have digital ICTs made informational privacy one of the most obvious and pressing issues in computer ethics? The question is crucial[3] and deceptively simple.

According to one of the most widely accepted explanations, digital ICTs exacerbate old problems concerning informational privacy because of the dramatic increase in their data *Processing* capacities, in the speed (or *Pace*) at which they can process data, and in the *Quantity* and *Quality* of data that they can collect, record and manage. This can be referred to as the 2P2Q hypothesis.

The trouble with any approach sharing the 2P2Q hypothesis is that it concentrates only on obvious and yet secondary effects of the *digital* revolution, and that it does so from a "continuist" philosophy of technology (more on this later). It thus fails to account for the equally important fact that digital ICTs are also responsible

[1] Especially in the US, see Garfinkel (2000).

[2] Froomkin (2000) still provides a valuable review.

[3] See for example Johnson (2001), Bynum and Rogerson (2004) and Tavani (2003).

both for a potential *increase* in some kinds of informational privacy and, above all, for a radical *change* in its overall nature. ICTs are more redrawing rather than erasing the boundaries of informational privacy. A few examples may help to illustrate the point. Consider

- the "remotization" of information management, such as the ordinary phenomenon of booking, banking or shopping online;
- the growth of anonymous, indirect or non- personal interactions. According to a recent surveyby Freever (a mobile-services firm, http://www.freever.com) 45% of Britons had lied about their location by text message; this is privacy as well;
- the much faster and more widespread revisability, volatility and fragility of digital data. Personal records can be upgraded or erased at the stroke of a key, destroyed by viruses in a matter of seconds, or become virtually unavailable with every change in technological standards, whereas we are still able to reconstruct whole family trees thanks to parish documents that have survived for centuries; or
- the various technologies that enable users to encrypt, firewall or protect information (e.g. with passwords or PIN).

In each case, it seems that digital ICTs allow both the erosion of informational privacy and its protection. The following, colourful episode is indicative: "Hong Kong businessmen, for example, once did not dare to leave their mobile phones switched on while visiting sleazy Macau, because the change in ringing tone could betray them. After the ringing tone for Macau was changed to sound like Hong Kong's, however, they could safely leave their phones on, and roaming revenues soared." (*The Economist*, December 2nd 2004).

The new challenges posed by digital ICTs are not only a matter of "more of the same". They have their roots in a radical and unprecedented transformation in the very nature (ontology) of the informational environment, of the informational agents[4] embedded in it and of their interactions. As will be argued in this article, understanding this ontological transformation provides a better explanation that is not only consistent with the 2P2Q hypothesis—now to be interpreted as a mere secondary effect of a far more fundamental change—but also closer to the kernel of the privacy problem in the information society.

4.3 Informational Privacy as a Function of Ontological Friction

Imagine a model of a limited (region of the) info sphere, represented by four students (our set of interactive, informational agents) living in the same house (our limited environment). Intuitively, given a certain amount of available information

[4] For a precise definition of agent see Floridi and Sanders (2004b).

(which can be treated as a constant and hence disregarded), the larger the *informational gap* among the agents, the less they know about each other, the more private their lives can be.

The informational gap is a function of the degree of accessibility of personal data. In our example, there will be more or less informational privacy depending on whether the students are allowed, e.g., to have their own rooms and lock their doors. Other relevant conditions are easily imaginable (individual fridges, telephone lines in each room, separate entrances, etc.).

Accessibility, in its turn, is an epistemic factor that depends on the ontological features of the infosphere, i.e. on the nature of the specific agents, of the specific environment in which they are embedded and of the specific interactions implementable in that environment by those agents. If the walls in the house are few and thin and all the students have excellent hearing, the degree of accessibility is increased, the informational gap is reduced and informational privacy is more difficult to obtain and protect. The love life of the students may be deeply affected by the Japanese-style house they have chosen to share.

The ontological features of the infosphere determine a specific degree of "ontological friction" regulating the information flow within the system. "Ontological friction" refers here to the forces that oppose the information flow within (a region of) the infosphere, and hence (as a coefficient) to the amount of work required for a certain kind of agent to obtain information (also, but not only) about other agents in a given environment, e.g. by establishing and maintaining channels of communication and by overcoming obstacles in the flow of information such as distance, noise, lack of resources (especially time and memory), amount and complexity of the data to be processed, etc.

Of course, the informational affordances and constraints provided by an environment are such only in relation to agents with specific informational capacities. In our model, brick walls provide much higher "ontological friction" for the flow of acoustic information than a paper-thin partition, but this is irrelevant if the students are deaf. More realistically, the debate on privacy issues in connection with the design of office spaces (from private offices to panel-based open plan office systems, to completely open working environments, see Becker and Sims (2000)) offers a significant example of the relevance of varying degrees of ontological friction in social contexts.

We are now ready to formulate a qualitative sort of equation, which will be needed to analyze the relation between digital ICTs and informational privacy. Given a certain amount of personal information available in (a region of) the infosphere I, the lower the ontological friction in I, the higher the accessibility of personal information about the agents embedded in I, the smaller the informational gap among them, and the lower the level of informational privacy implementable about each of them. Put simply, informational privacy is a function of the ontological friction in the infosphere. It follows that any factor affecting the latter will also affect the former. The factors in question can vary and may concern more or less temporary or reversible changes in the environment (imagine three of our students living in a tent during a holiday, while the fourth is left home alone) or in the agents (e.g., two of our students change their behaviour because the other two have quarreled).

Because of their "data superconductivity", ICTs are well-known for being among the most influential factors that affect the ontological friction in the infosphere.[5] A crucial difference between old and new ICTs is how they affect it.

4.4 Ontological Friction and the Difference Between Old and New ICTs

In the past, ICTs have always tended to reduce what agents considered the normal degree of ontological friction in their environment. This already held true for the invention of the alphabet or the diffusion of printing. Photography and the rise of the daily press were no exceptions. One can easily sympathize with nineteenth century concerns about the impact on individuals' informational privacy of "[r]ecent inventions and business methods [...], [i]nstantaneous photographs and newspaper enterprise [...] and numerous mechanical devices" (Warren and Brandeis 1890).

All this does not mean that, throughout history, informational privacy has constantly decreased in relation to the invention and spreading of ever more powerful ICTs. This would be a simplistic and mistaken inference. As emphasized above, changes in the nature both of the environment and of the agents play a pivotal role as well, so the actual ontological friction, and hence the corresponding degree of informational privacy in a region of the infosphere, are the result of a fine balance among several factors. Most notably, during the nineteenth and the twentieth centuries, following the industrial revolution, the social phenomenon of the new metropolis counteracted the effects of the latest ICTs, as urban environments fostered a type of informational privacy based on *anonymity.*[6] This is the sort of privacy enjoyed by a leaf in the forest, still inconceivable nowadays in rural settings or small villages. In the same period in which Warren and Brandeis were working on their classic article, the Edinburgh of Dr. Jekyll[7] and the London of Sherlock Holmes[8] already provided increasing opportunities for informational privacy through anonymity, despite the recent availability of new technologies.

Old ICTs have always tended to reduce the ontological friction in the infosphere because they enhance or augment the agents embedded in it. To understand why, consider the appliances available in our students' house.

---

[5] For a similar point see Moor (1997), who writes "When information is computerized, it is *greased* to slide easily and quickly to many ports of call" (p. 27).

[6] Anonymity is defined here as the unavailability of personal information, or the "noncoordinability of traits in a given respect", according to Wallace (1999).

[7] Stevenson's *The Strange Case of Dr Jekyll and Mr Hyde* was first published in 1886.

[8] Doyle's *A Study in Scarlet* was first published in 1887.

Some appliances—e.g. a drill, a vacuum cleaner or a food mixer—are tools that enhance their users, exactly like an artificial limb. Tele-ICTs (e.g. the telescope, the telegraph, the radio, the telephone or the television) are enhancing in this sense. Some other appliances—e.g. a dishwasher, a washing machine or a refrigerator—are robots that augment their users insofar as well-specified tasks can be delegated to them, at least partially. Recording ICTs (e.g. the alphabet and the various writing and printing technologies, the tape or video recorder) are augmenting in this sense.

Enhancing and augmenting ICTs have converged and become bundled together. The Watergate scandal and Nixon's resignation would have been impossible without them. But whether kept separate or packaged together, old ICTs have always shared the fundamental feature of facilitating the information flow in the infosphere by increasingly empowering the agents embedded in it. This "agent-oriented" trend in old, predigital[9] ICTs is well represented by dystopian views of informationally omnipotent agents, able to overcome any ontological friction, to control every aspect of the information flow, to acquire any personal data and hence to implement the ultimate surveillance system, thus destroying all informational privacy, "the dearest of our possessions".

Now, according to a "continuist" interpretation of technological changes, digital ICTs should be treated as just one more instance of well-known, enhancing or augmenting ICTs. But then—the reasoning goes—if there is no radical difference between old and new (i.e. digital) ICTs, it is reasonable to argue that the latter cause increasing problems for informational privacy merely because they are orders of magnitude more powerful than past technologies in enhancing or augmenting agents in the infosphere. All past ICTs have tended to reduce the ontological friction in the infosphere by enhancing or augmenting the agents inhabiting it, but digital ICTs are no exception, so the 2P2Q explanation is correct. Orwell's "Big Brother" is readily associated with the ultimate database.

Although the continuist 2P2Q hypothesis is reasonable and intuitive, it overlooks the essence of the problem. In theory, ontological friction can both be reduced and increased. We have seen how the emergence of the urban environment actually produced more anonymity, and hence more ontological friction and more informational privacy. The difference between old and new ICTs is that the former tended to reduce informational privacy, whereas the latter can also increase it. This is because the former tended to enhance or augment the agents involved more and more, whereas the latter can also change the very nature of the infosphere (that is, of the environment itself, of the agents embedded in it and of their interactions). The 2P2Q explanation misses a fundamental difference between old and new ICTs: the former are enhancing or augmenting whereas the latter are best understood as re-ontologizing technologies, an important distinction that needs to be analyzed in some detail.

[9] Orwell's *1984*, first published in 1949, contains no reference to computers or digital machines.

4.5 Digital ICTs as Re-ontologizing Technologies

Our model and a bit of science fiction will help to introduce the new concept of re-ontologization.[10]

Suppose that all the walls and the furniture in our students' house are transformed into perfectly transparent glass. Assuming our students have good sight, this will drastically reduce the ontological friction in the system. Imagine next that the students are transformed into proficient mind-readers and telepathists. Any informational privacy in this sort of Bentham's PanOpticon will become virtually impossible. The thought experiment illustrates how radical modifications in the very nature (a re-ontologization) of the infosphere can dramatically change the conditions of possibility of informational privacy.

The influence exercised by the new digital ICTs on the infosphere can now be analyzed in terms of its re-ontologization. Schematically, one can distinguish five fundamental trends.

1. The digitization of the informational environment. This is the most obvious way in which the new ICTs have re-ontologized the infosphere. The transition from analogue to digital data is very familiar and requires no explanation, but perhaps a brief comment may not go amiss. In their second study on information storage and flows, Lyman and Varian (2003) write that "Print, film, magnetic, and optical storage media produced about 5 exabytes of new information in 2002. Ninety-two percent of the new information was stored on magnetic media, mostly in hard disks. [...] Five exabytes of information is equivalent in size to the information contained in 37,000 new libraries the size of the Library of Congress book collections" (Lyman and Varian 2003). Although the production of analogue data is still increasing, the infosphere is fast becoming progressively more digital.
2. The homogenization of the processor and the processed. The re-ontologization of the infosphere has also been caused by the fundamental convergence between digital resources and digital tools. The ontology of the information technologies available (e.g. software, databases, communication protocols etc.) is now the same as (and hence fully compatible with) the ontology of their objects. This was one of Turing's most consequential intuitions: in the re- ontologized infosphere, there is no longer any substantial difference between the processor and the processed and the digital deals effortlessly and seamlessly with the digital. This potentially eliminates one of the most long-standing bottlenecks in the infosphere, a major source of ontological friction. The increasing computerization of artefacts (from the cash machine to the fridge, from the car to the building, from one's underwear to a book, cf. the current debate on privacy and RFID[11]) and of whole social environments (the phenomenon of "Ubiquitous Computing" or

[10] The neologism is constructed following the word "re- engineering" ("to design and construct anew").

[11] Radio Frequency IDentification, a method of storing and remotely retrieving data using tags or transponders.

"Ambient Intelligence"[12]) reminds us that soon it will be difficult to understand what life was in predigital times.

3. The evolution of new informational agents. This change concerns the emergence of artificial and hybrid agents (i.e. partly artificial and partly human; consider the group of our students as a single agent, equipped with digital cameras, laptops, palm pilots, mobiles, a wireless network, digital TVs, DVDs, CD players, etc.). These new artificial agents share the same ontology with their environment and can operate in it with much more freedom and control. This is where digital ICTs can be mistaken for mere augmenting technologies. Arguably, the infosphere will be progressively populated by artificial or hybrid agents, to which other (not necessarily human) agents will be able to delegate tasks and decisions. It is to be expected that the moral status of such agents will become an ever more challenging issue.[13]
4. The informationalization of interactions. In the re-ontologized infosphere populated by ontologically-equal entities and agents, where there is no ontological difference between processors and processed, interactions become equally digital. They are all interpretable as "read/write" (i.e., access/alter) activities, with "execute" the remaining type of process.
5. The mutation of old agents into informational agents. Finally, by re-ontologizing the infosphere, digital ICTs have also brought to light the intrinsically informational nature of human agents. This is not equivalent to saying that our students in the house have digital alter egos, some Messrs. Hydes represented by their @s, blogs and https. This trivial point only encourages us to mistake digital ICTs for merely enhancing technologies. The informational nature of agents should not be confused with a "data shadow"[14] either. The more radical change, brought about by the re-ontologization of the infosphere, has been the disclosure of human agents as informational entities among other informational entities, in the following sense.

Recall the distinction between enhancing and augmenting appliances. The switches and dials of the former are interfaces meant to plug in the appliance to the user's body ergonomically. The data and control panels of augmenting appliances are instead interfaces between different possible worlds: on the one hand there is the human user's Umwelt,[15] and on the other hand there are the dynamic, watery, soapy, hot and dark world of the dishwasher; the equally watery, soapy, hot and dark but also spinning world of the washing machine; or the still, aseptic, soapless, cold and potentially luminous world of the refrigerator. These robots can be successful

[12] Coroama et al. (2004), Bohn et al. (2004) and Brey (2005) offer an ethical evaluation of privacy-related issues in Ambient Intelligence environments. For a technically informative and balanced assessment see also Gow (2005).

[13] The issue of artificial morality is analyzed in Floridi and Sanders (2004b).

[14] The term is introduced by Westin (1968) to describe digital profile generated from data concerning a user's habits online.

[15] The outer world, or reality, as it affects the agent inhabiting it.

because they have their environments "wrapped" and tailored around their capacities, not vice versa. Imagine our students trying to build a droid like C3PO capable of washing their dishes in the sink exactly in the same way as they would.

Computers and digital ICTs are not augmenting or empowering in the sense just explained. They are ontologizing devices because they engineer environments that the user is then enabled to enter through (possibly friendly) gateways. So, whilst a dishwasher interface is a panel through which the machine enters into the user's world, a computer interface is a gate through which a user can be telepresent in the info- sphere (Floridi 2005). This simple but fundamental difference underlies the many spatial metaphors of "cyberspace", "virtual reality", "being online", "surfing the web", "gateway" and so forth. The re-ontologization of the infosphere, just sketched, has been causing an epochal, unprecedented migration of humanity from its Umwelt to the info- sphere itself. Inside it, humans are informational agents among other informational (possibly artificial) agents. They operate in an environment that is friendlier to "digital creatures". They have the ontological status of informational entities. And as digital immigrants are replaced by digital natives, the latter may come to appreciate that there is no ontological difference between infosphere and Umwelt, only a difference of levels of abstractions (Floridi and Sanders 2004a, b).

4.6 Informational Privacy in the Re-ontologized Infosphere

To summarize, so far it has been argued that informational privacy is a function of the ontological friction in the infosphere. Many factors can affect the latter, including, most importantly, technological innovations and social developments. Old ICTs affected the ontological friction in the infosphere mainly by enhancing or augmenting the agents embedded into it; therefore, they tended to decrease the degree of informational privacy possible within the infosphere. On the contrary, digital ICTs affect the ontological friction in the infosphere most significantly by re-ontologizing it; therefore, not only can they both decrease and protect informational privacy but, most importantly, they can also alter its nature and hence our understanding and appreciation of it.

Framing the revolutionary nature of digital ICTs in this ontological way offers several advantages. The first can be highlighted immediately: the ontological hypothesis is perfectly consistent with the 2P2Q hypothesis, since the re-ontologization of the infosphere explains why digital ICTs are so successful, in terms of the quantity, quality and speed at which they can variously process their data. It follows that the ontological hypothesis can inherit whatever explanatory benefits are carried by the 2P2Q hypothesis.

Four other advantages can be listed here but each of them requires a more detailed analysis: (1) contrary to the 2P2Q hypothesis, the new approach explains why digital ICTs can also enhance informational privacy, although (2) there is still a sense in which the information society provides less protection for informational privacy than the industrial society did. Above all, (3) the ontological hypothesis provides the right frame within which to assess con- temporary interpretations of informational privacy and (4) can indicate how we might wish to proceed in the future in order to protect informational privacy in the newly re-ontologized infosphere. Let us consider each point in turn.

4.7 Empowering the Informational Agent

In the re-ontologized infosphere, any informational agent has an increased power not only to gather and process personal data, but also to control and protect them. Recall that the digital now deals with the digital effortlessly. The phenomenon cuts both ways. It has led not only to a huge expansion in the flow of personal information being recorded, processed and exploited, but also to a large increase in the types and levels of control that agents can exercise on their personal data. And while there is only a certain amount of personal data that an agent may care to protect, the potential growth of digital means and measures to control their life-cycle does not seem to have a foreseeable limit. If privacy is the right of individuals (being these single persons, groups, or institutions) to control the life-cycle (especially the generation, access, recording and usage) of their information and determine for themselves when, how, and to what extent their information is processed by others, then one must agree that digital ICTs may enhance as well as hinder the possibility of enforcing such right.

At their point of generation, digital ICTs can foster the protection of personal data, e.g. by means of encryption, anonymization, password-encoding, firewalling, specifically devised protocols or services, and, in the case of externally captured data, warning systems.

At their point of storage, legislation, such as the Data Protection Directive passed by the EU in 1995, guarantees that no ontological friction, already removed by digital ICTs, is surreptitiously reintroduced to prevent agents from coming to know about the existence of personal data records, and from accessing them, checking their accuracy, correcting or upgrading them or demanding their erasure.

And at their point of exploitation—especially through data-mining, sharing, matching and merging—digital ICTs could help agents to control and regulate the usage of their data by facilitating the identification and regulation of the relevant users involved.

At each of these three stages, solutions to the problem of protecting informational privacy can be not only self-regulatory and legislative but also technological, not least because informational privacy infringements can more easily be identified and redressed also thanks to digital ICTs.

All this is not to say that we are inevitably moving towards an idyllic scenario in which our PETs (Privacy Enhancing Technologies) will fully protect our private lives and information against harmful PITs (Privacy Intruding Technologies). Such optimism is unjustified. But it does mean that digital ICTs can already provide some means to counterbalance the risks and challenges that they represent for informational privacy, and hence that no fatalistic pessimism is justified either. Digital ICTs do not necessarily erode informational privacy; they can also enhance and protect it. A good example is provided by the P3P (Platform for Privacy Preferences) initiative of the W3C (World Wide Web Consortium, see http://www.w3.org/P3P/).

4.8 The Return of the (Digital) Community

Because digital ICTs are radically modifying our informational environments, ourselves and our interactions, it would be naive to expect that informational privacy in the future will mean exactly what it meant in the industrial Western world in the middle of the last century.

Previously, we saw that, between the end of the nineteenth and the beginning of the twentieth century, the ontological friction in the infosphere, actually reduced by old ICTs, was nevertheless increased by social conditions favouring anonymity and hence a new form of informational privacy. In this respect, the diffusion of digital ICTs has finally brought to completion the process begun with the invention of printing. We are back into the now digital community, where anonymity can no longer be taken for granted, and hence where the decrease in ontological friction caused by old and new ICTs can have all its full-blown effects on informational privacy. In Britain, for example, public places are constantly monitored by 1.5 m CCTV systems, with the result that the average citizen is recorded 300 times a day (The Economist, (Jan 23rd 2003). The digital ICTs that allowed terrorists to communicate undisturbed over the Internet were also responsible for the identification of the London bombers in a matter of hours. Likewise, mobile phones are increasingly useful as forensic evidence in trials. In Britain, cell site analysis (a form of triangulation that estimates the location of a mobile phone when it is used) helped disprove Ian Huntley's alibi and convict him for the murdering of Holly Wells and Jessica Chapman. Sherlock Holmes has the means to fight Mr. Hyde.

How serious and dangerous is it to live in a glassy infosphere? Human agents tend to be acquainted with different environments that have varying degrees of ontological friction and hence to be rather good at adapting themselves accordingly. As with other forms of fine equilibria, it is hard to identify, for all agents in any environments, a common, lowest threshold of ontological friction below which human life becomes increasingly unpleasant and ultimately unbearable. It is clear, however, that a particular threshold has been overcome when the agents are willing to employ resources, run risks or expend energy to restore it, e.g. by building a higher fence, by renouncing a desired service, or by investing time in revising a customer profile. On the other hand, different agents have different degrees of sensitivity. One needs to remember that several factors (character, culture, upbringing, past experiences etc.) make each agent a unique individual. To one person, a neighbour capable of seeing one's garbage in the garden may seem an unbearable breach of their privacy, which it is worth any expenditure and effort to restore; to another person, living in the same room with several other family members may feel entirely unproblematic. Human agents can adapt to very low levels of onto- logical friction. Virginia Woolf's essay on Montaigne discusses the lack of ontological friction that characterizes public figures in public contexts. Politicians and actors are used to environments were privacy is a rare commodity. Likewise, people involved in "Big Brother" (but "Truman Show" would be a more appropriate label) programmes show a remarkable capacity to adapt to settings where any ontological friction between them and the public is systematically reduced, apparently for the sake of entertainment. In far more tragic and realistic contexts, prisoners in concentration camps are subject to extreme duress due to both intended and unavoidable rarefaction of ontological friction (Levi 1959).

The information society has revised the threshold of ontological friction and therefore provides a different sense in which its citizens appreciate their informational privacy. Your supermarket knows exactly what you like, but so did the owner of the grocery where your grandparents used to shop. Your bank has detailed records of all your visits and of your financial situation, but how exactly is this different from the old service? A phone company could analyze and transform the call data collected for billing purposes into a detailed subscriber profile: social network (names and addresses of colleagues friends or relatives called), possible nationality (types of international calls), times when one is likely to be at home and hence working patterns, financial profile (expenditure) and so forth. Put together the data from the supermarket, the bank and the phone company, and inferences of all sorts could be drawn for one's credit rating. Yet so they could be and were in Alexandre Dumas' *The Count of Monte Cristo* (1844). Some steps forward into the information society are really steps back into a small community and, admittedly, the claustrophobic atmosphere that may characterize it.

In the early stages in the history of the Web, roughly when Netscape was synonymous with browser, users believed that being online meant being entirely anonymous. A networked computer was like Gyges' ring in Plato's *Republic* (359b–360d): it made one invisible, unaccountable and therefore potentially less responsible, socially speaking. Turing would certainly have appreciated the (at the time) popular comic strip in which a dog, typing an email on a computer, confessed to another dog that "when you are on the Internet nobody can guess who you are". Nowadays, the strip is not funny anymore, only outdated. Cookies, monitoring software and malware (malicious software, such as spyware) have made people realize that the screen in front of them is not a shield for their privacy or Harry Potter's invisibility cloth, but a window on their lives online, through which virtually anything could be seen. They expect web sites to monitor and record their activities and do not even mind for what purpose. They accept that being online is one of the less private things in life.[16]The screen is a monitor and is monitoring you.

[16]"The best long-term assessment of public attitudes toward privacy is provided by Columbia's Alan Westin, who has conducted a series of polls over the last 30 years on this issue. On average, he finds that one quarter of the American public cares deeply about keeping personal information secret, one quarter doesn't care much at all, and roughly half are in the middle, wanting to know

A few years ago, a journalist at The Economist ran an experiment (The Economist, December 16th 1999). He asked a private investigator, "Sam", to show what information it was possible to gather about someone. The journalist himself was to be the subject of the experiment. The country was Britain, the place where the journalist lived. The journalist provided Sam with only his first and last names. Sam was told not to use "any real skulduggery (surveillance, going through her domestic rubbish, phone-tapping, hacking, that sort of thing)". The conclusion? By using several databases and various ICTs, "Without even talking to anyone who knows me, Sam [...] had found out quite a bit about me. He had a reasonable idea of my personal finances—the value of my house, my salary and the amount outstanding on my mortgage. He knew my address, my phone number, my partner's name, a former partner's name, my mother's name and address, and the names of three other people who had lived in my house. He had 'found' my employer. He also had the names and addresses of four people who had been directors of a company with me. He knew my neighbours' names."

Shocking? Yes, in the anonymous industrial society, but not really in the pre-industrial village before it, or in the information society after it. In Guarcino, a small village south of Rome of roughly a thousand people, everybody knows everything about everybody else, "vita, morte e miracoli", "life, death and miracles", as they say in Italian. There is very little ontological friction provided by anonymity so there is very little informational privacy in that respect. A difference with the information society is that we have seen that the latter has the digital means to protect what the small village must necessarily forfeit.

There are of course many other dissimilarities. As Paul Oldfield has rightly stressed,[17] the comparison between today's information society and the small community of the past, where "everybody knows everything", must be taken with more than a pinch of salt. History may repeat itself, yet never too monotonously. Small communities had a high degree of intra-community transparency (like a shared house) but a low degree of inter-community transparency (they were not like the Big Brother house, visible to outside viewers). So in those communities, the breaches of privacy were reciprocal, yet there were few breaches of privacy across the boundary of the community. This is quite different from today's information society, where there can be very little transparency within the communities we live or work in (we hardly know our neighbours, and our fellow-workers have their privacy rigorously protected), yet data-miners, hackers and institutions can be very well informed about us. Breaches of privacy from outside are common. What is more, we do not even know whether they know our business. On the other hand, part of the value of this comparison lies in the size of the community taken into consideration. A special trait of the information society is precisely its lack of boundaries, its global nature.

---

more about the benefits, safeguards, and risks before providing information. Customer behaviour in the marketplace – where many people freely provide personal information in exchange for various offers and benefits – seems to bear out this conclusion" Walker (2000).

[17] Private communication. The rest of this section is largely based on comments sent to me by Paul Oldfield,

We live in a single infosphere, which has no "outside" and where intra- and inter-community relations are more difficult to distinguish. The types of invasion of privacy are quite different too. In the small community, breaches of privacy might shame or discredit you. Interestingly, Augustine usually speaks of privacy in relation to the topic of intercourse in married couples, and he always associates it to secrecy and secrecy to shame or embarrassment. Or they might disclose your real identity or character (more on this later). Things that were private became public knowledge. In the information society, such breaches involve unauthorized collection of information, not necessarily its publication. Things that are private may not become public at all; they may be just accessed and used by privileged others. The small community also had its own self-regulations for limiting breaches of privacy. Everyone knew that they were as subject to scrutiny as everyone else, and this set an unspoken limit on their enthusiasm for intruding into others' affairs.

4.9 Assessing Theories of Privacy

Once it is acknowledged that digital ICTs have re-ontologized the infosphere, it becomes easier to assess the available theories of informational privacy and its moral value.

Two theories are particularly popular: the *reductionist interpretation* and the *ownership-based interpretation*.

The reductionist interpretation argues that the value of informational privacy rests on a variety of undesirable consequences that may be caused by its breach, either personally (e.g. distress) or socially (e.g. unfairness). Informational privacy is a utility, also in the sense of providing an essential condition of possibility of good human interactions, e.g. by pre- serving human dignity or by providing political checks and balances.

The ownership-based interpretation argues that informational privacy needs to be respected because of each person's rights to bodily security and property (where "property of x" is classically understood as the right to exclusive use of x). A person is said to own his or her information (information about him- or herself)—recall Virginia Woolf's "infinitely the dearest of our possessions"—and therefore to be entitled to control its whole life-cycle, from generation to erasure.[18]

The two approaches are not incompatible, but they stress different aspects of informational privacy. One is more oriented towards a consequentialist assessment of privacy protection or violation. The other is more oriented towards a "natural rights" understanding of the concept of privacy itself, in terms of private or intellectual property. Unsurprisingly, they both compare privacy breach to a

[18] The debate on the ownership-based interpretation developed in the seventies, see Scanlon (1975) and Rachels (1975), who criticize Thomson (1975), who supported an interpretation of the right to privacy as being based on property rights.

trespass[19] or unauthorized invasion of, or intrusion in, a space or sphere of personal information, whose accessibility and usage ought to be fully controlled by its owner and hence kept private. A typical example is provided by the border-crossing model of informational privacy developed by Gary T. Marx since the late nineties (see now Marx 2005).

The reductionist interpretation is not entirely satisfactory. Defending the need for respect for informational privacy in view of the potential misuse of the information acquired is certainly reasonable, especially from a consequentialist perspective, but it may be inconsistent with pursuing and furthering social interests and welfare. For, although it is obvious that even some public personal information may need to be protected—e.g. against profiling or unrestrained electronic surveillance—it remains unclear, on a purely reductionist basis, whether a society devoid of any informational privacy may not be a better society, with a higher, common welfare.[20] It has been argued, for example, that the defence of informational privacy in the home may actually be used as a subterfuge to hide the dark side of privacy: domestic abuse, neglect or mistreatment. Precisely because of reductionist-only considerations, even in democratic societies such as the UK and the US, it tends to be acknowledged that the right to informational privacy can be overridden when other concerns and priorities, including business needs, public safety and national security, become more pressing. All this is despite the fact that article 12 of The Universal Declaration of Human Rights clearly indicates that "No one shall be subjected to arbitrary interference with his privacy, family, home or correspondence, nor to attacks upon his honour and reputation. Everyone has the right to the protection of the law against such interference or attacks."

The ownership-based interpretation also falls short of being entirely satisfactory. Three problems are worth highlighting here:

1. the issue of informational contamination under- mining passive informational privacy; this is the unwilling acquisition of information or data (e.g. mere noise) imposed on someone by some external source. Brainwashing may not occur often, but junkmail, or the case of a person chatting loudly on a mobile near us, are unfortunately very com- mon experiences of passive privacy breach, yet no informational ownership seems to be violated;
2. the issue of informational privacy in public con- texts; informational privacy is often exercised in public spaces, that is, in spaces which are not only socially and physically public—a street, a car park, a pub—but also informationally public—anyone can see the newspaper one buys, the bus one takes, the T-shirt one wears, the drink one is ordering (Patton 2000). How could a CCTV system be a breach

[19] See Spinello (2005) for a recent assessment of the use of the trespassing analogy in computer-ethical and legal contexts. Charles Ess has pointed out to me that comparative studies have shown such spatial metaphors to be popular only in Western contexts.

[20] Moor (1997) infers from this that informational privacy is not a core value, i.e. a value that "all normal humans and cultures need for survival", but then other values he lists as "core" are not really so in his sense, e.g. happiness and freedom. According to Moor, privacy is also intrinsically valuable, while being the expression of the core value represented by security.

of someone's privacy if the agent is accessing a space which is public in all possible senses anyway? and

3. the metaphorical and imprecise use of the concept of "information ownership", which cannot quite explain the lossless acquisition (or usage) of information: contrary to other things that one owns, one's personal information is not lost when acquired by someone else. Analyses of privacy based on "ownership" of an "informational space" are metaphorical twice over.

4.10 The Ontological Interpretation of Informational Privacy and its Value

Both the reductionist and the ownership-based interpretation fail to acknowledge the radical change brought about by digital ICTs. They belong to an industrial culture of material goods and of manufacturing/trading relations. They are overstretched when trying to cope with the new challenges offered by an informational culture of services and usability. Warren and Brandeis (1890) had already realized this limit very insightfully: "where the value of the production [of some information] is found not in the right to take the profits arising from publication, but in the peace of mind or the relief afforded by the ability to prevent any publication at all, *it is difficult to regard the right as one of property, in the common acceptation of the term*" (p. 25, emphasis added).

More than a century later, in the same way as the digital revolution is best understood as a fundamental re-ontologization of the infosphere, informational privacy requires an equally radical re-interpretation, one that takes into account the essentially informational nature of human beings and of their operations as informational social agents.

Such re-interpretation is achieved by considering each person as constituted by his or her information, and hence by understanding a breach of one's informational privacy as a form of aggression towards one's personal identity.

The following passage by Marcel Proust, though admittedly referring to the social construction of the individual, helps to conveys the idea of a person as an informational entity: "But then, even in the most insignificant details of our daily life, none of us can be said to constitute a material whole, which is identical for everyone, and need only be turned up like a page in an account-book or the record of a will; our social personality is created by the thoughts of other people. Even the simple act which we describe as 'seeing someone we know' is, to some extent, an intellectual process. We pack the physical outline of the creature we see with all the ideas we have already formed about him, and in the complete picture of him which we compose in our minds those ideas have certainly the principal place. In the end they come to fill out so completely the curve of his checks, to follow so exactly the line of his nose, they blend so harmoniously in the sound of his voice that these seem to be no more than a transparent envelope, so that each time we see the face or hear the voice it is our own ideas of him which we recognize and to which we listen." (*Remembrance of Things Past—Swann's Way*).

The ontological interpretation is consistent with the fact that digital ICTs can both erode and rein- force informational privacy, and hence that a positive effort needs to be made in order to support not only PET but also "poietic" (i.e. constructive) applications, which may allow users to design, shape and maintain their identities as informational agents (Floridi and Sanders 2005). The information flow needs some friction in order to keep firm the distinction between the multiagent system (the society) and the identity of the agents (the individuals) constituting it. Any society (even a utopian one) in which no informational privacy is possible is one in which no personal identity can be maintained and hence no welfare can be achieved, social welfare being only the sum of the individuals' involved. The total "transparency" of the infosphere that may be advocated by some reductionists—recall the example of the glassy house and of our mentally super-enhanced students—achieves the protection of society only by erasing all personal identity and individuality, a "final solution" for sure, but hardly one that the individuals them- selves, constituting the society so protected, would be happy to embrace freely.

The advantage of the ontological interpretation over the reductionist one is then that consequentialist concerns may override respect for informational privacy, whereas the ontological interpretation, by equating its protection to the protection of personal identity, considers it a fundamental and inalienable right,[21] so that, by default, the presumption should always be in favour of its respect. As we shall see, this is not to say that informational privacy is never negotiable in any degree.

Looking at the nature of a person as being constituted by that person's information allows one to understand the right to informational privacy as a right to personal immunity from unknown, undesired or unintentional changes in one's own identity as an informational entity, either actively—collecting, storing, reproducing, manipulating etc. one's information amounts now to stages in cloning and breeding someone's personal identity—or passively—as breaching one's informational privacy may now consist in forcing someone to acquire unwanted data, thus altering her or his nature as an informational entity without consent.[22] The first difficulty facing the ownership-based interpretation is thus avoided: in either case, the ontological interpretation suggests that there is no difference between one's informational sphere and one's personal identity. "You are your information", so anything done to your information is done to you, not to your belongings. The right to informational privacy (both in the active and in the passive sense just seen) shields one's personal identity. This is why informational privacy is extremely valuable and ought to be respected.

Heuristically, violations of informational privacy are now more fruitfully compared to a digital kid- napping rather than trespassing: the observed is moved to an

[21] For a different view see Volkman (2003).

[22] This view is close to the interpretation of privacy in terms of protection of human dignity defended by Bloustein (1964).

observer's local space of observation (a space which is remote for the observed), unwillingly and possibly unknowingly. What is abducted is personal information and no actual removal is in question, but a cloning of the relevant piece of personal information. Yet the cloned information is not a "space" that belongs to the observed and which has been trespassed; it is part of the observed herself, or better something that (at least partly) constitutes the observed for what she or he is. It is a Doppelganger, as Richard Avedon described it once, when speaking of his photograph of Henry Kissinger ("Is it just a shadow representation of a man? Or is it closer to a doppel-ganger, a likeness with its own life, an inexact twin whose afterlife may overcome and replace the original?"). A further advantage, in this change of perspective, is that it becomes possible to dispose of the false dichotomy qualifying informational privacy in public or in private contexts. Insofar as a piece of information constitutes an agent, it does so context- independently and that is why the observed may wish to preserve her integrity and uniqueness as an informational entity, even when she is in an entirely public place. After all, trespassing makes no sense in a public space, but kidnapping is a crime independently of where it is committed. The second problem affecting the ownership-based interpretation is also solved.

As for the third problem, one may still argue that an agent "owns" his or her information, yet no longer in the metaphorical sense seen above, but in the precise sense in which an agent is her or his information. "My" in "my information" is not the same "my" as in "my car" but rather the same "my" as in "my body" or "my feelings": it expresses a sense of constitutive belonging, not of external ownership, a sense in which my body, my feelings and my information are part of me but are not my (legal) pos- sessions. It is worth quoting Warren and Brandeis (1890) once again: "[...] the protection afforded to thoughts, sentiments, and emotions [...] is merely an instance of the enforcement of the more general right of the individual to be let alone. It is like the right not to be assaulted or beaten, the right not to be imprisoned, the right not to be maliciously persecuted, the right not to be defamed [or, the right not to be kidnapped, my addition]. In each of these rights [...] there inheres the quality of being owned or possessed and [...] there may be some propriety in speaking of those rights as property. But, obviously, they bear little resemblance to what is ordinarily comprehended under that term. *The principle* [...] *is in reality not the principle of private propriety but that of inviolate personality* (p. 31, emphasis added) [...] *the right to privacy, as part of the more general right to the immunity of the person,* [is] *the right to one's personality* (p. 33, emphasis added)".

This ontological conception has started being appreciated by more advanced information societies where identity theft is the fastest growing white-collar offence. Informational privacy is the other side of identity theft, to the point that, ironically, for every person whose identity has been stolen (around 10 m Americans are victims annually) there is another person (the thief) whose identity has been "enhanced".

Recent problems affecting Google and its privacy policy convey a similar picture. As Kevin Bankston, staff attorney at the Electronic Frontier Foundation, remarks "Your search history shows your associations, beliefs, perhaps your medical problems. *The things you Google for define you.* [...] data that's practically a printout of what's going on in your brain: What you are thinking of buying, who you talk to, what you talk about" (quoted in Mills 2005, emphasis added).

As anticipated, the ontological interpretation reshapes some of the assumptions behind our still "industrial" conception of informational privacy. Three examples are indicative of this transition.

If personal information is finally acknowledged to be a constitutive part of someone's personal identity and individuality, then 1 day it may become strictly illegal to trade in some kinds of personal information, exactly as it is illegal to trade in human organs (including one's own) or slaves. The problem of child pornography may also be revisited in light of an ontological interpretation of informational privacy. At the same time, one might relax one's attitude towards some kinds of "dead personal information" that, like "dead pieces of oneself", are not really or no longer constitutive of oneself. One should not sell one's kidney, but can certainly sell one's hair or be rewarded for giving blood. Recall the experiment of the journalist at The Economist. Very little of what Sam had discovered could be considered ontologically constitutive of the person in question. We are constantly leaving behind a trail of personal data, pretty much in the same sense in which we are losing a huge trail of dead cells. The fact that nowadays digital ICTs allow our data trails to be recorded, monitored, processed and used for social, political or commercial purposes is a strong reminder of our informational nature as individuals and might be seen as a new level of ecologism, as an increase in what is recycled and a decrease in what is wasted.

At the moment, all this is just speculation and in the future it will probably be a matter of fine adjustments of ethical sensibilities, but the third Geneva Convention (1949) already provides a clear test of what might be considered "dead personal information": a prisoner of war need only give his or her name, rank, date of birth, and serial number and no form of coercion may be inflicted on him or her to secure any further information, of any kind. If we were all considered "prisoners of the information society", our informational privacy would be well protected and yet there would still be some personal data that would be perfectly fine to share with any other agent, even hostile ones.

A further issue that might be illuminated by the ontological interpretation is that of confidentiality. The sharing of private information with someone, implicitly or explicitly, is based on a relation of pro- found trust that joins together the agents involved. This coupling is achieved by allowing the agents to be partly constituted, ontologically, by the same information. Visually, the informational identities of the agents involved now overlap, at least partially, as in a Venn diagram. The union of the agents forms a single unity, a supra-agent. Precisely because entering into a new supra-agent is a delicate and risky operation, care should be exercised before "melding" oneself with other individuals by sharing personal information or its source i.e. common experiences. Confidentiality is a bond that is hard and slow to forge properly, yet resilient to many external forces when finally in place, as the supra-agent is stronger than the constitutive agents themselves. Relatives, friends, classmates, fellows, colleagues, comrades, companions, partners, team-mates, spouses and so forth may all have experienced the nature of such a bond, the stronger taste of a "we". But it is also a bond very brittle and difficult to restore when it comes to betrayal, since the disclosure, deliberate or unintentional, of some personal information in violation of confidence can entirely and irrecoverably destroy the privacy of the new, supra-agent born out of the joining agents, by painfully tearing them apart. We shall return to the topic of trust and confidentiality at the end of this article.

A third and final issue can be touched upon rather briefly, as it was already mentioned above: the ontological interpretation stresses that informational privacy is also a matter of construction of one's own informational identity. The right to be let alone is also the right to be allowed to experiment with one's own life, to start again, without having records that mummify one's personal identity forever, taking away from the individual the power to mould it. Everyday, a person may wish to build a different, possibly better, "I". We never stop becoming our- selves, so protecting a person's informational privacy also means allowing that person the freedom to change, ontologically.[23]

4.11 Informational Privacy, Personal Identity and Biometrics

On September 12, 1560 the young Montaigne attended the public trial of Arnaud du Tilh, an impostor who was sentenced to death for having faked his identity. Many acquaintances and family members, including the wife Bertrande, had been convinced for a long while that he was Martin Guerre, returned home after many years of absence. Only when the real Martin Guerre came home was Arnaud's actual identity finally ascertained.

Had Martin Guerre always been able to protect his personal information, Arnaud du Tilh would have been unable to steal his identity. Clearly, the more one's informational privacy is protected the more one's personal identity can be safeguarded. This new qualitative equation is a direct consequence of the onto- logical interpretation. Personal identity also depends on informational privacy. The difficulty facing our contemporary society is how to combine the new equation with the other equation, introduced above, according to which informational privacy is a function of the ontological friction in the infosphere. Ideally, one would like to reap all the benefits from

(a) the highest level of information flow; and hence from
(b) the lowest level of ontological friction; while enjoying
(c) the highest level of informational privacy protection; and hence

[23] In this sense, Johnson (2001) seems to be right in considering informational privacy an essential element in an individual's autonomy. Moor (1997), referring to a previous edition of Johnson (2001), disagrees.

(d) the highest level of personal identity protection.

The problem is that (a) and (d) seem incompatible: facilitate and increase the information flow through digital ICTs and the protection of one's personal identity is bound to come under increasing pressure. You cannot have an identity without having an identikit. Or so it seems, until one realizes that the information flowing in (a) consists of all sorts of data, including arbitrary data about oneself (e.g. a name and surname) that are actually shareable, whereas the information required to protect (d) can be ontic data, that is, data constituting someone (e.g. someone's DNA) that are hardly sharable by nature.[24]

Personal identity is the weakest link and most delicate element in our problem. Even nowadays, personal identity is regularly protected and authenticated by means of some arbitrary data, randomly or conventionally attached to the bearer/user, like a mere label: a name, an address, a Social Security number, a bank account, a credit card number, a driving licence number, a PIN and so forth. Each label in the list has no ontologically constitutive link with its bearer; it is merely associated with someone's identity and can easily be detached from it without affecting it. The rest is a mere consequence of this "detachability". The more the ontological friction in the infosphere decreases, the swifter these detached labels can flow around, and the easier it becomes to grab and steal them and use them for illegal purposes. Arnaud du Tilh had stolen a name and a profile and succeeded in impersonating Martin Guerre for many years in a rather small village, within a community that knew him well, fooling even Martin's wife, apparently. Eliminate all personal interactions and identity theft becomes the easiest thing in the world.

A quick and dirty way to fix the problem would be to clog the infosphere by slowing down the information flow. Building some traffic calming device, as it were. It seems the sort of policy popular among some IT officers and bank managers, keen on not allowing this or that operation for security reasons, for example. However, as with all counter-revolutionary or anti-historical approaches, "resistance is futile": trying to withstand the evolution of the infosphere only harms current users and, in the long run, fails to deliver an effective solution.

A much better approach is to ensure that the ontological friction keeps decreasing, thus benefiting all the inhabitants of the infosphere, while safe- guarding personal identity by data that are not arbitrary labels about, but rather constitutive traits of, the person in question. Arnaud du Tilh and Martin Guerre looked very similar, yet this was as far as biometrics went in the sixteenth century. Today, biometric digital ICTs are increasingly used to authenticate a person's identity by measuring the person's physiological traits—such as fingerprints, eye retinas and irises, voice patterns, facial patterns, hand measurements or DNA sampling—or behavioral features, such as typing patterns. Since they also require the person to be identified to be physically present at the point-of-identification, biometric sys- tems provide a very reliable way of ensuring that the person is who the person claims to be; of course not always, and not infallibly—after all Montaigne used the extraordinary

[24] On the tripartite distinction between information as, about or for reality see Floridi (2004).

case of Martin Guerre to challenge human attempts ever to reach total certainty—but far more successfully than any arbitrary label can. It is a matter of degree.

All this is not to say that we should embrace bio- metrics as an unproblematic panacea. As Alterman (2003) has correctly shown, there are many risks and limits in the use of such technologies as well. But it is significant that digital ICTs, in their transformation of the information society into a digital community, are partly restoring, partly improving (see the case of Martin Guerre) that reliance on personal acquaintance that characterized relations of trust in any small town. By giving away some information, one can safeguard one's identity and hence one's informational privacy, while taking advantage of interactions that are personalized (through preferences derived from one's habits and behaviours) and customized (through preferences derived from one's expressed choices). In the digital community, you are a recognized individual, whose tastes, inclinations, habits, preferences etc. are known to the other agents, who can adapt their behaviour accordingly.

As for protecting the privacy of biometric data, again, no rosy picture should be painted, but if one applies the "Convention of Geneva" test, it seems that even the worst enemy could be allowed to authenticate someone's identity by measuring her fingerprints or his eye retinas. They seem to be personal data that is worth sacrificing in favour of the extra protection they can offer of one's personal identity and private life.

Once a cost/benefit analysis is taken into account, it makes sense to rely on authentication systems that do not lend themselves so easily to misuse. In the digital community, one is one's own information and can be (biometrically) recognized as oneself as one was in the small village. The case of Martin Guerre is there to remind us that mistakes are still possible. But their likelihood decreases dramatically the more biometric data one is willing to check. On this, Penelope can teach us a final lesson.

4.12 Conclusion

When Odysseus returns to Ithaca, he is identified four times. Argos, his old dog, is not fooled and recognizes him despite his disguise as a beggar. Then Eurycleia, his wet-nurse, while bathing him, recognizes him by a scar on his leg, which he had received from a boar when hunting. He then proves to be the only man capable of stringing Odysseus' bow. All these are biometric tests no Arnaud du Tilh would have passed. But then, Penelope is no Bertrande either. She does not rely on any "unique identifier" but finally tests Odysseus by asking Eurycleia to move the bed in their wedding-chamber. Odysseus protests that this is impossible: he himself had built the bed around a living olive tree, which is now one of its legs. This is a crucial piece of information that only Penelope and Odysseus ever shared. By naturally relying on it, Odysseus restores Penelope's full trust. She recognizes him as the real Odysseus not because of who he is or how he looks, but, onto- logically, because of the information that they have in common and that constitutes both of them as a couple. Through the sharing of this piece of information, identity is restored and the supra-agent is reunited. There is a line of continuity between the roots of the olive tree and the married couple. For Homer, their bond was homophrosyne (like-mindedness); to Shakespeare, it was the marriage of true minds. To us, it is informational privacy that admits no ontological friction.

Acknowledgements Previous versions of this paper were presented at the seminar Bridging Cultures: Computer Ethics, Culture, and ICT (NTNU, Trondheim, 6 June, 2005) and at the Sixth International Conference of Computer Ethics: Philosophical Enquiry (CEPE2005, University of Twente, 18 July, 2005). I wish to thank all the participants and especially Phil Brey, Charles Ess, Johnny Søraker, Bernd Carsten Stahl and May Thorseth and the anonymous referees of the journal for their comments. Paul Oldfield and Matteo Turilli read a final version and suggested several crucial improvements. None of them is responsible for any of its shortcomings.

References

Alterman, A. (2003). A piece of yourself: Ethical issues in biometric identification. *Ethics and Information Technology, 5*(3), 139–150.

Becker, F., & Sims, W. (2000). *Offices that work: Balancing cost, flexibility, and communication.* New York: Cornell University Inter- national Workplace Studies Program.

Bloustein, E. (1964). Privacy as an aspect of human dignity: An answer to dean Prosser. *New York University Law Review, 39*, 962–1007.

Bohn, J., Coroama, V., Langheinrich, M., Mattern, F., & Rohs, M. (2004). Social, economic, and ethical implications of ambient intelligence and ubiquitous computing. *Journal of Human and Ecological Risk Assessment, 10*(5), 763–786.

Brey, P. (2005). Freedom and privacy in ambient intelligence. In P. Brey, F. Grodzinsky, & L. Introna (Eds.), *Ethics of new information technology – Proceed- ings of the sixth international conference of computer ethics: Philosophical enquiry (Cepe2005)* (pp. 91–100). Enschende: CEPTES University of Twente.

Bynum, T. W., & Rogerson, S. (2004). *Computer ethics and professional responsibility*. Oxford: Blackwell.

Coroama, V., Bohn, J., & Mattern, F. (2004). Living in a smart environment – Implications for the coming ubiquitous information society. In *IEEE SMC 2004, the Hague, the Netherlands, October 10–13* (pp. 5633–5638).

Floridi, L. (1999). Information ethics: On the philosophical foundations of computer ethics. *Ethics and Information Technology, 1*(1), 37–56. Reprinted, with some modifications, in *The Ethicomp Journal*, 1(1), 2004, http://www.ccsr.cse.dmu.ac.uk/journal/articles/floridi_l_philosophical.pdf.

Floridi, L. (2004). Information. In L. Floridi (Ed.), *The Blackwell guide to the philosophy of computing and information* (pp. 40–61). Oxford: Blackwell.

Floridi, L. (2008). Information ethics. In J. van den Hoven & J. Weckert (Eds.), *Moral philosophy and information technology*. Cambridge: Cambridge University Press.

Floridi, L. (2005). Presence: From epistemic failure to successful Observability. *Presence: Teleoperators and Virtual Environments, 14*(6), 656–676.

Floridi, L., & Sanders, J. W. (2004a). The method of abstraction. In M. Negrotti (Ed.), *Yearbook of the artificial. Nature, culture and technology. Models in contemporary sciences* (pp. 177–220). Bern: Peter Lang.

Floridi, L., & Sanders, J. W. (2004b). On the morality of artificial agents. *Minds and Machines, 14*(3), 349–379.

Floridi, L., & Sanders, J. W. (2005). Internet ethics: The constructionist values of Homo Poieticus. In R. Cavalier (Ed.), *The impact of the Internet on our moral lives*. New York: SUNY.
Froomkin, A. M. (2000). The death of privacy? *Stanford Law Review, 52*, 1461–1543.
Garfinkel, S. (2000). *Database nation: The death of privacy in the 21st century*. Beijing: O'Reilly.
Gow, G. (2005). *Consumers and Privacy in Ubiquitous Network Societies – Background Paper*. Retrieved August 16, 2005, from http://www.itu.int/osg/spu/ni/ubiquitous/Papers/Privacy%20 background%20paper.pdf
Johnson, D. G. (2001). *Computer ethics* (3rd ed.). Upper Saddle River, NJ: Prentice-Hall.
Levi, P. (1959). *If this is a man*. London: Orion Press.
Lyman, P. & Varian, H. R.. (2003). *How much information?* Retrieved August 16, 2005, from http://www.sims.berkeley.edu/research/projects/how-much-info-2003/execsum.htm#summary
Marx, G. T. (2005). Some conceptual issues in the study of Borders and surveillance. In E. Zureik & M. B. Salter (Eds.), *Global surveillance and policing – Borders, security, identity*. Cullompton: Willan Publishing. chapter 2.
Mills, E. (2005). Google Balances Privacy, Reach. *C|Net News. com*. Retrieved August 30, 2005, from http://news.com.com/Google+balances+privacy%2C+reach/2100-1032_3-5787483.html
Moor, J. H. (1997). Towards a theory of privacy in the Informa- tion age. *ACM SIGCAS Computers and Society, 27*, 27–32.
Patton, J. W. (2000). Protecting privacy in public? Surveillance Technologies and the Value of Public Places. *Ethics and Information Technology, 2*(3), 181–187.
Rachels, J. (1975). Why privacy is important. *Philosophy and Public Affairs, 4*, 323–333.
Scanlon, T. (1975). Thomson on privacy. *Philosophy and Public Affairs, 4*, 315–322.
Spinello, R. A. (2005). Trespass and Kyosei in cyberspace. In R. A. Spinello & H. T. Tavani (Eds.), *Intellectual property rights in a networked world: Theory and practice*. Hershey, PA: Idea Group. chapter 8.
Tavani, H. T. (2003). *Ethics and technology: Ethical issues in an age of information and communication technology*. New York: Wiley.
The Economist December 2nd 2004 Your Cheating Phone.
The Economist December 16th 1999, Living in the Global Goldfish Bowl.
The Economist Jan 23rd 2003, SURVEY: THE INTERNET SOCIETY.
Thomson, J. (1975). The right to privacy. *Philosophy and Public Affairs, 4*, 295–314.
Turing, A. M. (1936). On computable numbers, with an Applica- tion to the Entscheidungsproblem. *Proceedings of the London Mathematical Society, 2*(42), 230–265.
Volkman, R. (2003). Privacy as life, liberty. *Property. Ethics and Information Technology, 5*(4), 199–210.
Walker, K. (2000). Where everybody knows your name: A pragmatic look at the costs of privacy and the benefits of information exchange. *Stanford Technology Law Review, 2*, 1–50.
Wallace, K. A. (1999). Anonymity. *Ethics and Information Tech- nology, 1*(1), 23–35.
Warren, S., & Brandeis, L. D. (1890). The right to privacy. *Harvard Law Review, 193*(4), 193–220.
Westin, A. F. (1968). *Privacy and freedom 1st*. New York: Atheneum.

Chapter 5
On the Meaning of Technology's Domination

Emanuele Severino

Abstract We keep talking about the increasing presence of technology in human life, but this discourse is always reduced to the description of a fact that is there, but may not be there, or may not have been there at all. Going beyond the ways in which philosophies of history (for instance, the Hegelian one or the Marxist one) have articulated a "necessary" development within history, this article foregrounds the inconsistency of this kind of "necessity", but, at the same time, shows how the reflection on dominating technology cannot be limited to a simple description of this phenomenon.

Keywords Metaphysics of technology · Technology's domination · Capitalism · Western culture · *Poiesis*

5.1 Introduction: Capitalism and Technology

Technology is now *dominant* in a human being's life. This statement has become commonplace. Scientific and philosophical research on such a domination has proliferated—yet, its meaning is still controversial. For a long time, my works have been showing the need of overcoming this controversy.

It should be noted that the capitalist production of wealth must be distinguished from technology, i.e., from what this production uses in order to achieve its aims, and that the capitalist globalization of today's world is destined to be substituted by what we may call "technological (or techno-scientific) globalization". Moreover,

E. Severino (✉)
Università Ca' Foscari Venezia, Università Vita-Salute San Raffaele, Accademia Nazionale dei Lincei, Rome, Italy

S. Chiodo, V. Schiaffonati (eds.), *Italian Philosophy of Technology*, Philosophy of Engineering and Technology 35,
https://doi.org/10.1007/978-3-030-54522-2_5

the kind of domination technology is destined to is the process that brings about the end not only of capitalism, but of all the big powers that compete to rule today's world, and that use technology in order to achieve their aims: states, planned economies, dictatorships and democracies, religions and all the relevant forms of the Western tradition (also those from the East, which, as I shall show, may be considered as the prehistory of the West). Capitalism is, among these powers, the most powerful, since it is mainly conceived in terms of embedding technology (and also since huge areas of worldwide techno-scientific research are funded by private capital). With the term "technology", I mean the system of subsystems (industrial, bureaucratic, medical, military, juridical, educational, related to mass media, etc.) that progressively adapt their parameters to those of modern science.

While lingering over these introductory remarks, it may be added that technology is destined to the radical domination of the world because it is *inevitable*—according to a meaning of inevitability that will be clarified—that all the powers that deploy technology in order to achieve their aims come to assume the enhancement of technology as their aim, and therefore stop being what they are.

I shall now proceed to clarify the authentic meaning of these statements—statements that, however, it is important to note, make reference to something that has nothing to do with "technocracy" as it is understood from Saint-Simon to Thorstein Veblen, up to the present day.

5.2 Aim and Means in Relation to Capitalism

First, I shall articulate the meaning of the relationship between aim and means. On this, it must be noted that every human action, from the simplest to the most complex one, is configured as it is configured (i.e., is what it is) because it has the aim that it has, and not others. The aim of an action determines the way in which it is constituted. It is a principle already known within Greek philosophical thought. If the aim of an action changes, then the action also becomes something different: only outwardly it remains what it used to be.

For instance, since the aim of the capitalist endeavour, and of capitalism in general, is the growth of private profit, capitalism would not be capitalism anymore if, for any reason, it was forced to assume an aim different from the one that belongs to it. (Given the relevance of this example for the issues I am going to develop, it should be noted that, in the production-consumption cycle, which, according to marginalism, is a process where consumption tends to be fashioned as the aim of the economic cycle, consumption, on the contrary, is actually the means that capitalism uses in order to increase profit, since a product that is not consumable is not saleable.) But what capitalism is heading towards is, as I shall show, a transformation where the original aim of capitalism is substituted by a different aim.

This transformation is not a simple "fact". "Facts" happen, but, according to today's dominant culture, they could not have happened, and may not happen. This transformation is not a simple "fact" and this is to be related to the circumstance that technology's proceeding towards the domination of the world is also not a simple "fact", but, as I have started to say, something inevitable, a "destiny". Even the philosophical-scientific thought, when acknowledging that technology is now *dominant* in a human being's life, considers this domination a "fact". Yet, things stand otherwise.

5.3 Conflicts Between Powers That Deploy Technology

The powers that deploy technology today in order to achieve their aims are mutually *conflicting*. In wide areas of the world (the United States, Europe, India, Japan, etc.), capitalism coexists with democracy (but it coexisted, and still coexists, also with non-democratic forms of social life). Yet, the aim of democracy is different from that of capitalism. Even if there are different forms of capitalism, the aim they share is the growth of *private profit*, whereas the aim of democracy is citizens' freedom and equality. Freedom of economic endeavour is, of course, one of the freedoms that democracy wants to protect, but democracy cannot allow the growth of private profit to be the ultimate aim of society. Likewise, capitalism, in its own essence, and despite its claims, cannot allow for the aspiration to individual freedom and equality as the ultimate aim of society to reduce the growth of profit beyond a certain limit. The more or less visible contrast among these two powers is permanent. The same goes for the conflict between capitalism and Christianity (or Islam). If, as Max Weber argues, the Protestant ethics helped the formation of the "spirit of capitalism", according to such an ethics, it is assured that worldly success (i.e., capitalist accumulation of wealth) is a *sign* of being on the road to eternal beatitude: worldly success is thus not the ultimate aim of life, since the ultimate aim is indeed this very beatitude. The Catholic Church acknowledges that, in comparison with the planned economy of socialism, capitalism has a greater power of producing wealth, but, at the same time, admonishes capitalism not to consider profit as its ultimate aim, but, rather, the "common good" of society.

I could develop further examples of the conflicting nature of the relationship between the powers that deploy technology today in order to achieve their aims. Yet, it is now important to outline, even if briefly, what happens within the relationship between these powers and the technology they deploy in order to prevail over the other powers. Nowadays, techno-science is the most powerful tool. (In the past, other tools were the most powerful: for instance, physical power together with courage or astuteness; and, according to religions such as Christianity, prayer is still the most powerful tool.) But any of these powers makes use of a part of the technological power that is at our disposal on earth today, and, if it wants to prevail over the others, or at least not to succumb, then it is forced non only to protect the survival of this part, but also to strengthen and extend it. And it is forced to *keep* doing it, because the other powers that do not want to succumb act likewise. Thus, inevitably, any fixed level of equality is continuously overcome—also because of the different interpretation that each of these powers attributes to the equality achieved. Moreover, this continuous strengthening of the part of technology that each power manages requires not to be impeded and weakened by the aim that each of them pursues, i.e., by the direction that every power gives to its own means in order to achieve its own aim. Weakening the means entails compromising the achievement of the aim.

5.4 Whoever Wins, Loses

Since this conflict is a matter of life and death, what we encounter in this conflict is an extreme way of prevailing over the antagonist. Since it is inevitable that the powers that own a smaller part of technological power will succumb, it is likewise inevitable that the most powerful ones, which own a basically equal part, will survive (where the part of technology that is owned also includes the capacity of managing it). Now, since the conflict goes on, it is inevitable that one of these powers, in order to prevail, decides to assume as its *unique* aim the strengthening of the technological means it deploys, therefore relinquishing the will of achieving its original aim, by which it is defined. Which entails that, conditions and power being equal (considering the equality resulting from the defeat of the powers that own a smaller part of technology), this *decision* makes this power prevails over those powers that keep having as their aim, *besides* the strengthening of their own technological means, *also* the aim that originally defines them, which therefore reduces the means devoted to achieving this technological aim. These powers lose, and the power that makes that decision wins. But does it really win?

"It", it is said, wins. But the power that wins is no longer the power that struggled in order to achieve its original aim. By making that *decision*, it has indeed abandoned this original aim. And, as I have said, since an action is what it is because of the aim it has, then the winner is a *different* kind of agent, which has assumed as its ultimate aim the strengthening of the technological apparatus that this kind of agent possesses. This may happen even if, at least at the beginning, the struggling powers do not realise what is going on.

5.5 The USA, the USSR and the Inevitability of Technological Domination

This outline of the result of the conflict between the powers that today try, in different ways, to lead the world is not the result of abstract considerations. (Yet, it is a prediction—and I shall be back on the meaning of this prediction.) Today it is still the case that the two strongest powers, which more or less directly coordinate all the others, are the United States and Russia. They are the strongest ones because they have a nuclear armament that makes them invincible in comparison with every other enemy except for themselves—and each of the two can destroy the other by being destroyed in the process. Nuclear power is essential for technology, because, if today these two powers are still going to deploy technology for their own aims, then technology, whoever is making use of it, has its own aim: unlimitedly increasing its own capacity of achieving aims. And this is possible only if this kind of increase is not threatened or even destroyed by hostile powers. Nuclear power is able, more than any other power, to avoid these threats and destruction. The United States coordinate the smaller nuclear powers, such as Israel, the UK, France, India; Russia coordinates powers such as China, Iran, North Korea. Even if the Chinese economy may move closer to or surpass that of the United States, it will never be able to close the gap at nuclear level.

After the collapse of the USSR, the United States and Russia are nowadays two different forms of capitalism: as their aim, they have a different form of capitalism (even if the difference is based on identity), and are aware that it is not possible to impose this form on the whole planet. In order to achieve this, they would have to descend into conflict and destroy each other (but not even the USA and the USSR would descend into proper conflict, as I used to say even during the most acute phases of the Cold War). Beyond the more or less long-lasting agreements, they are in conflict anyway. That is, they are in a situation in which there is the possibility of making the above-mentioned *decision* (paragraph 4), i.e., one of these two powers, in order to prevail, could decide to assume as its *unique* aim the strengthening of the technological means it deploys, relinquishing the will of achieving its own original aim. And it is inevitable that, within this situation of conflict, this decision is made, and that this kind of strengthening, whilst including the strengthening of nuclear arsenals, excludes nuclear conflict, which, by destroying the world, would not strengthen, but destroy, technological means. If that decision is made by only one of the two powers, then this power will inevitably prevail. Yet, if it is made by them both, then the aim of them both will be the undefined strengthening of the technological apparatus, i.e., their difference, and therefore the root of their conflict, will stop.

By making that decision, the two most powerful owners of technology—and therefore the groups that each of them gathers—are destined to become technology's servants, and the same goes for the less strong powers (religions, democracy and politics, the values of Western and Eastern traditions, the pressure of poor societies over the rich ones), which, even if mutually conflicting, are now becoming the means of the worldwide capitalist economy. But, because of that decision, it is the worldwide capitalist economy that in turn becomes the means that the worldwide technological apparatus deploys in order to increase its power indefinitely. If, at this point, this apparatus keeps being managed by both the powers that made that decision, then their being different and their maintaining their own autonomies extend their conflict without any reason except for the very will of being different, and this conflict weakens their powers. Thus, it is inevitable that it is precisely their will to power what overcomes their difference, until they constitute a unique power, whose aim (the increase of power) becomes the aim of the whole planet. This is the time of the worldwide government ruled by the *pax technica*.

5.6 Production, Destruction and Becoming Something Else

But what has been said so far, beyond being brief, is still extremely far from the authentic meaning of technology's destiny of world domination. Indeed, what is the essence of technology? And what do "destination", "power", "production", "destruction", and "world" mean? What is the relationship between these concepts and those of "aim" and "means"? And what is the conceptual status of the prediction I have proposed—also considering critiques of the presumption of extending the concept of prediction to the reflection on social events, concept which, instead, should find its proper application only within the natural sciences?

Power is said to be the capacity of producing and destroying things. We never reflect enough on the fact that, since the beginning of human life on earth, and during the whole of human history, the word "thing" has had a constant meaning or character—which is such even if, in the different languages, it is expressed in different ways, and related to different semantic contexts (for instance, in ancient Greek it is *prágma* and *chréma*, in Latin *res*, in German *Ding*). The constant meaning implied in the different ways of intending and expressing the "thing" is "being what *becomes something else* than what it is and *has become from something else* than what it was". In some uses of this word, it is also said that God is a "thing", and God is not understood as becoming something else and from something else. God is eternal and immutable not by being a thing, but by being a specific way of being a thing that does not involve all the other things. In *Genesis*, things, made by God, come from God, and John's *Apocalypse* says that they will become destroyed things. Production and destruction are things becoming from something else and becoming something else. Among the powers of production and destruction there is also the will (both human and divine). Will means making things become something else and from something else, i.e., transforming (or trying to transform) the world. Thus, the will is founded on the conviction that things—that are transformable—exist, and for their own nature are open to becoming something else and from something else through the will.

5.7 *Poiesis*

Since its very beginning in Greek culture, philosophical thought has been taking to extremes the meaning of that "something else" from which things come, and which then become. That is, it has been taking to extremes the meaning of production and destruction. This something else is the "*non-being*". Things become *non-being* (where they "are not anymore") and from *their* non-being (where they "are not yet"). When a building is built, *a lot* of what constitutes it pre-exists it (the place where it is going to rise, the building materials, the architect's design, the workers who are going to build it, etc.): a lot pre-exists it, *but not all* that it will be once built. If all it is pre-existed it, then it could not be built, since it would already exist. Thus, a thing can be produced only if, before being produced, a part, an aspect of it, and moreover its specific and coherent configuration, *are not*, i.e., are *nothing*, nothing *at all*. Likewise, a thing can be destroyed only if a part, an aspect of it, and moreover its specific and coherent configuration, *are not* anymore, are nothing by now, nothing at all—even if when, for instance, a building is destroyed, a lot of what constituted it still remains.

For the first time, philosophy evokes the radical meaning of *nothing*. That is, for the first time, it considers the thing as *being*. Then, "production" (*poíesis*) manifests itself as "the cause (*aitía*) that makes anything pass from non-being to being" (*ek toû mè óntos eis tò ón*), as Plato says in the *Symposium* (205 b-c). Thus, destruction is the "cause" that makes anything pass from being to non-being. Since it is a being, every being is this extreme form of becoming something else and from something else. By stating the existence of an immutable Being, metaphysics indeed states its immutability not because it is a being, but, also in this case, because it is a certain being that is privileged in comparison with every other being. And metaphysics also states, in some of its forms, that the totality of beings, and therefore every aspect and part of every being, comes from its absolute non-being.

Thus, the will starts manifesting as the cause that, having a certain aim, decides to make this aim pass from non-being (i.e., from its being only "ideally present") to being. Therefore, it is founded on the *conviction* that *beings* (what is transformable) *exist* and are in themselves *open* to going from non-being to being and vice versa because of certain causes, among which the will itself. And philosophical thought quite soon was persuaded that the existence of beings, together with that openness of them, is an unquestionable obviousness, an undeniable truth. Western history, even when it does not realise it or pretends not to be interested in this kind of suggestions, evolves starting from this conviction as a foundation, which is now worldwide.

Every form of will is now founded on this conviction. Even the will of technology. In the above-mentioned passage of the *Symposium*, Plato indeed notices that also "technology" (*téchne*) is "production", *poíesis*, "cause" of the oscillation of things between being and non-being. Even if it may seem implausible, Western history, as well as human history—and not only the history of thoughts, but also the history of works—evolves from the grounds that Greek philosophy founded. Even the claim according to which this evolution is implausible starts from these grounds. Technology is the most radical and legitimate way in which original philosophical thought manifests itself.

5.8 Philosophy and Technology

Yet, philosophy proceeds because the initial faith in things becoming something else allows human will to survive, but also makes the extreme danger appear. At a certain moment in this process, becoming something else means dying. The disorientation caused by death produces myth, which is above all a reassurance on the destiny of the dead. But the danger is too severe: even if, only after several millennia, human beings are not happy anymore with the mythical shelter, and want to look at death with a knowledge that cannot be denied, which philosophy is meant to be since its very beginning. Since its origin, philosophy seeks indeed to state the "necessity" (*chreôn*, Anaximander says) that all beings *pre-exist* within the immutable and divine "Principle" (*arché*) from which they come, and to which all of them, through their deaths, go back.

Since the very beginning, and during the whole Western tradition, *tradition*, philosophical thought is therefore an *antinomy*. On the one hand, it states that beings come (partly or completely) from their non-being, and go back to it—and that this coming from is an unquestionable obviousness. On the other hand, starting, in turn, from a conceptual structure that intends to manifest itself as an unquestionable obviousness, philosophical thought states that beings come from the immutable and divine "Principle" that initially contains them, and, when they get destroyed, they go back to it (even if in various ways). This configuration of the philosophical matrix of Western history has decisive consequences on the process that is destined to bring about technology's domination. Precisely for this reason, every reflection on technology that overlooks this configuration fails to understand the authentic meaning of technology. Thus, even the techno-scientific understanding of technology fails to understand it.

Since philosophy begins as the will to be immutable (incontrovertible) knowledge, which sets the immutable Being at the beginning and end of the becoming beings, philosophy qualifies itself as "science of truth", *epistéme tes aletheías*, i.e., manifestation (*alétheia*) of beings, since it is the capacity of *staying* (*–stéme*), imposing itself *on* (*epí*) any power that intends to remove it. Even modern science, for centuries, up to Einstein and even beyond, wants to have, more or less consciously, this character. But *if*, after the development of the philosophical tradition, *we try to look beyond the surface of philosophy's last two centuries*, then it is apparent that philosophy demonstrates, *within its essential substratum*, the impossibility of any *epistéme tes aletheías*. The conviction that beings come from their non-being and go back to it (the conviction that first envelops the very *epistéme* of the truth) mandatorily requires the nonexistence of any immutable knowledge and any immutable Being.

Indeed, the Immutable is the Law everything must adapt to: everything, and therefore also the totality of the past and the future, i.e., any being that is already nothing and that is still nothing. But that nothing that adapts itself to the law of the Immutable is not nothing, but a being. By affirming the existence of the Immutable, we come to think that *this nothing* (of the past and of the future) *is a being*: we come to think, against our own intentions, that those coming from nothing and going back to nothing, whose undeniable existences we are sure of anyway, do not exist. It is denied the very obviousness that nonetheless is affirmed. By denying the existence of any Immutable (in order not to identify nothing and being, and therefore not to deny the processes of becoming something else and from something else), the *antinomy* stated in the penultimate paragraph is resolved.

5.9 On the *de facto* Technological Domination

Until science and technology cannot think of this antinomy and its solution, i.e., cannot hear the voice of what I called the "*essential substratum* of philosophy's last two centuries", their domination is only a "fact" from the point of view of the *epistéme* of the truth: they prevail as a matter of fact, and not rightfully; their power is a prevarication. Indeed, science and technology in themselves cannot have the capacity of responding to the religious-theological-metaphysical knowledge of the Western tradition. The latter admonishes techno-science not to overcome the Limits traced by the Law, which are constituted by the *epistéme* of the truth, as well as by the immutable Being affirmed by it—the Law of the totality of beings. Thus, the admonishment makes reference to the undefined growth of technological power, which is determined by the conflict between the powers that deploy technology. In the meantime, science has in turn abandoned the will to be the *epistéme* of the truth, and it establishes itself, on the one hand, as a *specialisation* that is not interested in the totality of beings, and, on the other hand, and even in mathematics, as a hypothetical and falsifiable knowledge, i.e., a form of knowledge that does not have the authority and the right to oppose the great Western tradition of knowledge. (The two sides are mutually implied. Since the beginning of philosophical thought, the will to know the undisputable truth implies that the latter is the truth of the totality of beings, so that it cannot happen that this truth, by being the truth of a part of the Whole, is denied by the truth of the other parts.)

Besides, the *antinomy* I am referring to, and which is at the heart of the Western cultural tradition, can easily be obscured, since the faith *in both* the opposites that constitute it is too strong: on the one hand, the faith that the process of beings coming from their not being yet, as well as their process of heading towards not being anymore, is an absolute and undeniable obviousness; on the other hand, the opposite faith that saves us from the anguish caused by death, by stating that beings are eternally contained within the divine Principle of the Whole, and that they go back to this Principle through their destruction and death. The faith in these two opposites is too strong to allow us the power of quickly freeing ourselves from such a Principle. A faith that, moreover, considers itself as the supreme undeniable truth.

5.10 The Essential Substratum of Our Times

Thus, the process that brings techno-science to prevail over the other powers manifests itself as *inevitable* (as mentioned in paragraphs 3 and 4) only when technology gets unified with the *essential substratum* of philosophy's last two centuries (see paragraph 8), i.e., with the dimension that is able to be, *within the faith in the obviousness of becoming something else*, the "undeniable truth" that can survive after God's death, as well as after the death of the *epistéme* of the truth. Such a substratum is not only the essential substratum of philosophy's last two centuries, but it is the essential substratum of our era. It is necessary to go deeply within it, and illuminate it. Its inhabitants are just a few, and are misunderstood by culture worldwide (See Nietzsche 1968, 2005; Leopardi 1923, 1983, 2013), or completely ignored (See Gentile 1922, 1923).

The essential substratum is at the core of an atmosphere that is spreading out on earth: the denial of any definitive "Truth" and any immutable "Reality". Some examples follow. I have already mentioned the hypothetical and specialised character science has assumed (the beginning of non-Euclidean geometry and quantum physics' principle of indeterminacy are aspects of this character). To this, we should add the process that, since the Second World War, has brought about the fall of the European totalitarian states in Germany and Italy, as well as, after the end of the Cold War, in Russia—i.e., the fall of those states that posit themselves as immutable structures, in that they are true. Furthermore, I shall mention the end of the kind of democracy that wants to conform to the values of the Western tradition, and that is substituted by procedural democracy, which overlooks those values and in which what decides the value of social plans is the majority's vote. And the downfall of "natural right" and "natural morality" (where what is "natural" is given, once again, by traditional values), which are progressively substituted by "positive right" and "positive morality" (where what is "positive" is established by historically winning social groups, which establish as right and moral the values they believe in, and for which they have fought). And European art's distance from the "natural" model of beauty, through the imposition of "abstract" art, atonal music, and "fragment". And the abandonment, in advanced societies, of the social and individual behaviours of the Western tradition. Finally (but the list could go on considerably), the process that, as I have demonstrated in paragraphs 1–4, is leading techno-science to prevail over all the powers that intend to deploy it in order to achieve their aims. This is the atmosphere of which the last two centuries' philosophical essential substratum is the core—meaning that, without it, such an atmosphere would be the more or less conscious disorientation of those who left the ancient certainties without knowing yet what to hold on to.

5.11 The Essential Substratum and Technology

But the power of the substratum's voice cannot keep being unheard: it is "destined" to get heard by the powers that deploy technology, as well as by technology itself. By hearing and understanding this voice, technology radically transforms its own essence, and reaches the greatest form of the will to power. That voice tells technology (see paragraph 8) that the obviousness of becoming nothing and from nothing excludes the existence of any Immutable, i.e., of any absolute Law that has the right to limit the technological will to increase its own power infinitely by prevailing over any other form of will to power. Thus, that voice makes technology *actually* powerful, and at the highest level. A man who does not know that he is rich lives as poor; but he lives as rich when he is reliably told to be rich. Likewise, the powers that deploy technology, until they address that voice, more or less explicitly consider their own aims as absolute truths, and therefore resist the process in which they are sacrificed in order to strengthen the technological means and prevail over the opposing aims. But this resistance stops when these powers start doubting, and finally deny, the unquestionable truth of their own aims. The end of the USSR may indeed be interpreted as the process in which the unquestionability of Marxist philosophy started being doubted (i.e., one of the main ways in which the Western *epistéme* of the truth has been constituted), and Marxism started being considered as an obstacle to the efficiency and competitiveness of the techno-scientific apparatus it deployed and wanted to lead; and what prevailed was the will to strengthen that apparatus against the one managed by the United States. This process is not yet due to the conscious hearing of the voice of our era's essential substratum, but it is an important premonition of it.

5.12 Prediction and Contradiction

The statement according to which technology is "destined" to lead the world is also a *prediction*, which cannot be a scientific one, since, given the way in which science constitutes itself today—given its present incapability of hearing the voice of the essential substratum—, science cannot know anything about the history that, starting from the birth of philosophical thought and the *epistéme* of the truth, leads to the essential substratum of our era, and therefore cannot know anything about the relationship between such a substratum and technology. But the fact that this is not a scientific prediction does not mean that it is a prophecy. On the contrary, it means that it has not only a content, but also a radicality, which are extraneous to the *present* form of scientific knowledge.

These are extraneous to the *present* form, but not to the form implied by such a knowledge when it gets unified with the essential substratum. The radicality of this union implies the radicality of the prediction this union leads to. The root of the West is the faith in the existence of the process of beings becoming something else, as well as in the will's capacity of making them become something else. This root—as I have mentioned in paragraph 8—denies that an Immutable being of any sort exists beyond the process of becoming, because any Immutable being (which is evoked starting from the acknowledgment of the reality of becoming) makes the nothing of the future and of the past a being, and therefore is a denial of the process of becoming something else (which the very metaphysical thought, by which the Immutable being is affirmed, acknowledges as an undeniable obviousness). The essence of both the destination and the prediction related to technology is indeed the manifestation of this double *contradiction*: the identification of being and nothing, and that acknowledgment of being something else, which is at the same time its own denial. And this contradiction is an essential *instability*. (Already within the faith in becoming something else, the contradiction is basically denied, at least because it is necessary that the other, which the being becomes, and from which the being becomes, is not identical respectively with what the being is before becoming something else and with what the being has become.) That is, if any human circumstance is a contradiction, then the instability of the contradiction belongs to the grounds which push towards overcoming the contradiction, and such a phenomenon is not a simple statistical-probabilistic tendency.

5.13 Statistical-Probabilistic Tendency

The instability determined by the manifestation and the denial of that double contradiction (and of the contradiction in general) is the necessary, but not sufficient, condition to overcome the contradiction itself. Hegel and Marx think that the presence of the contradiction, beyond being a necessary condition, is also a sufficient condition of its being overcome. This presence would determine the necessary historical development that leads to overcome the contradiction. But something like the *necessity* of this development belongs to the *epistéme* of the Immutable beings, and therefore disappears together with the disappearance of the latter. For this precise reason, both Hegel and Marx cannot know the contradiction implied by the affirmation of the existence of the truth and the immutable Beings.

After all, it is the presence of the contradiction that gives the greatest power to the tendency of overcoming it, and therefore to the prediction of its being overcome, i.e., to what I have called "inevitability" of its being overcome and technology's "destination" to domination. On the contrary, the simple statistical-probabilistic tendency, which overlooks the concept of contradiction, cannot exclude that precisely what is thought to happen more frequently is, vice versa, a contradiction, and that, vice versa, what happens less frequently is not a contradiction. (Which does not exclude that the prediction founded on the denial of that double contradiction may come to join the statistical-probabilistic prediction within the accomplished techno-scientific practice.) Techno-science avoids the contradiction, and yet, until now, in its present form, has not perceived the implication between prediction, tendency, and contradiction.

In Western history, the double contradiction due to the affirmation of the Immutable beings' existence is preeminent (even if it is generally hidden inside the philosophical substratum of our era), but it is not the only one. For instance, the *conflict* between the powers that deploy technology today constitutes a *united* system, where the powers *conflict* indeed. It is a "real opposition" that is *at the same time* a "logical contradiction" (to draw on a Kantian distinction, despite Kant's intentions). Furthermore, the contradiction of the conflict is an instability, a tendency, a destination, which allow to predict the overcoming of such a contradiction. Besides, also the claim according to which the conflict between those powers leads towards their subordination to technology is a set of predictions primarily founded on contradictions, and therefore on their own denial. Those powers are wills. And the fact that the will wants without obtaining is a contradiction that allows to predict the fight against the opposite wills that obstruct its very possibility of obtaining.

Even the thought of having at one's disposal a part of the technological apparatus without continuously empowering it in order to prevail is a contradiction. And, in the final conflict between the strongest powers, when they have at their disposal the decisive move—i.e., the decision to assume as *unique* aim the strengthening of the technological means, where the will abandons the achievement of its own original aim—, it is a contradiction the fact that one of them does not decide to do so, given that every power wants to prevail over the others.

Technology is the object of much blame today—and not only by the Western cultural tradition. However, all the reproaches come from other forms of that very will to power that animates technology. The most frequent reproach is that technology, left on its own, destroys human beings by turning into violence. Yet, the traditional Immutable beings, by making the process of becoming something else impossible, and therefore by destroying it, also destroy the very process of becoming something else by which human life is constituted—their violence is not less than that of technology. Moreover, human beings have always been considered as conscious powers that are able to organise means in order to achieve aims. (Even the mystic organises his/her own means in order to be surrounded and invaded by the divine.) But technology indeed consists in this organisation. That is, human beings have always been considered as technological beings, starting from the very root of the different and conflicting ways in which the Western tradition has interpreted them. Within technology, human beings find their own deepest being in a more coherent way than the way in which they find their own depth within God. In this sense, technology authentically considered—i.e., not technically or scientifically considered—cannot be "inhuman".

5.14 Faith in the Existence of Becoming Something Else

Ever since human beings have lived on earth, the process of things becoming something else and from something else has always been considered, more or less explicitly, as an absolutely undeniable obviousness. Yet, since it also precedes the process of becoming something else and from something else, it has often been recognised as *faith*. (If the obviousness is undeniable, faith cannot be undeniable, since it is precisely what can be denied.) This was the first indication that all I have said so far is still exceedingly far from the authentic essence of technology. So far, indeed, I have mentioned some of the aspects of the authentic essence of technology *by remaining within that faith*. And the burden on that faith is huge: it is the whole human history. It is decisive to decide upon how much that faith can carry. The "critical thinking" of our era, even if sharp, has never addressed this faith, because it is within it that this "thinking" has been developing too. In the space of this article, it is not possible to recall what I have said about the abyss of that faith in my works. Thus, this article leads toward the threshold of what is most important, albeit without crossing it. I will conclude by referring to my works about the ultimate grounds of technology, and therefore of our own civilization and human life on earth. But, first, I shall mention something about these aspects.

I have said (paragraph 13) that, when the voices of tradition blame the destructive violence of technology, they do not realise that also the Immutable and the Divine beings have this character. But, I add now, is it not true that the ultimate root of every destructiveness and violence, together with that of every "power", is the way in which, always and everywhere, the *thing* has always been considered and experienced? Is it not true that we should overturn Heraclitus' remark, and, rather than saying that the war is the mother of everything, we should say that the thing is the mother of every war? Since the thing is what becomes something else, it *divides itself* from itself: whatever is the other that it becomes, it kills what it used to be. And by *invading* the other that it becomes, it kills also this other. Even if the other is the non-being, the nothing of the thing, it, by becoming this other, makes what is nothing (within which the other was) a being. It already makes what is nothing a being, even before the Immutable beings do it. The other from which it becomes kills, in turn, itself. And, by invading the thing, it kills it. Thus, the violence of the *thing* is unconsciously affirmed by those who, like Heidegger, claim to "let the things be" (*Gelassenheit*).

Cross-References At this point, I should show what is most decisive, i.e., on the one hand, that the process of becoming something else is a content that does *not* manifest itself, is *not* experienced, observable, contestable; and, on the other hand, that stating that the *thing* is violence does not yet mean getting to the core of my works, where I show that the faith in the existence of the process of becoming something else is also the greatest *folly*, the greatest *mistake* (see Severino 2016, and also 1998, 2012, and 2017).

English translation by Simona Chiodo revised by Diletta De Cristofaro.

References

Gentile, G. (1922). In H. Wildon Carr (Ed.), *Theory of mind as pure act*. London: Macmillan.

Gentile, G. (1923). *Sistema di logica come teoria del conoscere*. Bari: Laterza.

Leopardi, G. (1923). In G. L. Bickersteth (Ed.), *The poems of Leopardi*. Cambridge: Cambridge University Press.

Leopardi, G. (1983). The moral essays (P. Creagh, Trans.). New York: Columbia University Press.

Leopardi, G. (2013). Zibaldone. In M. Caesar & F. D'Intino (Eds.), *The notebooks of Leopardi*. Londra, Penguin.

Nietzsche, F. (1968). In W. Kaufmann (Ed.), *The will to power*. New York: Vintage Book.

Nietzsche, F. (2005). Thus spoke Zarathustra. In G. Parkes (Ed.), *A book for everyone and nobody*. Oxford: Oxford University Press.

Severino, E. (1998). *Il destino della tecnica*. Rizzoli: Milano.

Severino, E. (2012). *Capitalismo senza futuro*. Rizzoli: Milano.

Severino, E. (2016). *The essence of nihilism*. London: Verso.

Severino, E. (2017). *Il tramonto della politica*. Rizzoli: Milano.

Chapter 6
Machine, Culture, and Robot

Carlo Sini

Abstract The human being is the product of instrumental work and the social action of language. Human life has always been characterized by technological evolution, which marks the difference between the human being and the natural world.

Keywords Sign · Voice · Language · Selfconsciousness · Robot

At the beginning of Vorträge und Aufsätze (Heidegger 1954), in his essay on technology, Martin Heidegger observes: "Everywhere we remain unfree and chained to technology, whether we passionately affirm or deny it" (Heidegger 1977, p. 4). According to Heidegger, the standard instrumental conception of technology prevents us from establishing a correct relation to it. Thus Heidegger distinguishes between human being and technology; the latter, if misunderstood, is a prison that keeps us in chains. Misunderstanding technology, that is conceiving it as an instrument or means, prevents us from articulating the correct relationship between human being and technology. This way of thinking is shared by a host of more recent views—such as those, which Heidegger would not endorse, which consider the human being to be a technological symbiote, namely a synthesis of biological and technical human being, whereby the constantly increasing dose of technology is a reason for concern (Longo 2003).

But what does "human being" mean? Kant had already raised this question. The reference, as much as the appeal to this world—human being—is always a problem, never a solution. Separating biological and technological human being is a non-sense. In the first place, it is clearly a technological human being who makes

C. Sini (✉)
Università degli Studi di Milano, Milano, Italy
e-mail: carlo.sini1@alice.it

S. Chiodo, V. Schiaffonati (eds.), *Italian Philosophy of Technology*, Philosophy of Engineering and Technology 35,
https://doi.org/10.1007/978-3-030-54522-2_6

the distinction. And, more generally, how does a human being without technology distinguish herself/himself from an animal? The Gospel says that the lilies of the field do not spin, and the birds of heaven do not work (Luke: 12, 27–28); only human beings work, properly speaking. This fact must be understood correctly. It is work, therefore, that makes us human. It is not human beings who produce work, but the opposite is the case: human beings are the ongoing product of labor.

How, therefore, should we conceive of work? In its fundamental and foundational meaning, work cannot merely be identified with the object of a particular science, such as economy; indeed, economy itself is the result of a certain kind of work. My thesis is that the human being, the living being who works, and therefore work itself, are the effect and product of instrumental action. Thus we find ourselves in the opposite situation to Heidegger's: technology is really an instrument and a means; this is how we should conceive it. But it is not easy to find reflections on the fundamental nature of the instrument which are adequate to the problem in question.

Today we know that the use of tools is very old. Already the hominids, millions of years ago, chipped stones with great skill. *Homo sapiens* is the heir of this operative "technique", not so much in his/her "sapient" configuration, but first and foremost as *homo habilis*. To become *sapiens*, we need an additional and very special tool, which is language. But what is "language"? Let us proceed one step at a time.

To clarify the nature of prelinguistic instrumental action, as I intend it here, let us consider the case of a simple instrument: the branch of a tree used as a stick (which also the chimpanzee, as is well known, occasionally employs). The stick is the exosomatic extension—as paleo-anthropologists would put it—of the arm. It prolongs and thus enhances the action of the arm, but it is not the arm. It comes from the world but is no longer simply or immediately a living part of it. The stick is precisely the medium; being in the middle, it joins the action of the living body (the *Leib*, as one would say in German) to one part of the surrounding world. This intermediate nature characterizes its effectiveness and its "virtue" (*aretè*).

Above all, the instrument-stick conveys to the agent the unconsciously "instrumental" nature of her/his body-thing (*Körper*, in German). The living and spontaneous action uses its own body (*Leib*) as an instrument (*Körper*), reducing it to its pure strength and inertia. It makes its way into the bush by reducing itself to a body-thing. The stick collects the action's inheritance, so to speak, while carrying out other invaluable functions as well. For example, it embodies the very purpose of one's instinctive action, making it self-translucent. In other words, the stick "educates" the arm. It shows what the arm should do in order to be more successful in its undertakings. But it also reveals the limit of the arm-stick's action by contrasting it with the inertia the branch-stick inherits from the world.

There follow two consequences of fundamental importance. First, the instrumental use of the stick teaches the agent about the "objective" inertia of the world. The whole question concerning the objectivity of knowledge—the idea of a world in itself endowed with characteristics independent of human agents—finds its root here. The "objectivity of the world" is originally a product of instrumental action and would not be conceivable without this basic instrument-world relationship. Secondly, precisely the inertia of the instrument, its structural limitations with respect to the world, the fact that the world-environment is also much other than "stick-able", points to the infinite improvability of the instrument and the invention of further tools. Whence the birth of technology's path, which draws from itself the stimulus to a progressive "complication" and endless transformation of instruments—a process which no power (or reason in general) can limit. Through tools, *homo* becomes more and more *habilis*. Her/his work reveals her/him to herself/himself as the agent she/he is (she/he reads her/his action in that of the instrument, from which she/he derives more "intelligent" activities and capacities); at the same time, she/he discovers what the world is like, what to do with it and in it—from her/his "human" point of view, that is.

It follows that work and knowledge are at the bottom one and the same: only work, thus understood, produces knowledge, while knowledge is always the product of work. Secondly, we understand that human beings are technical beings par excellence and essence: all unduly "humanistic" or "spiritualistic" chats, which depict Human being in himself/herself as a sort of solution to any problem, something to privilege and protect in its natural or mysterious essence, its special destinies and perhaps its derivation from an absolute and enigmatic "being", are lost.

However, one decisive piece of the story is missing. Why doesn't the chimpanzee work (or the birds of the sky and the lilies of the field)? His action is certainly intelligent, but it would make no sense to regard it as "cognitive". In fact, non-human living beings lack the instrument of instruments, the royal instrument, as the philosophers have often thought of it—from Aristotle to Bruno, Heidegger and, more recently, Foucault among others. The royal instrument is language. But again, what is language?

Conceiving language as an instrument and a medium is also the first part of the proposal here briefly outlined. In animal societies, Mead (1934) studied the function of so-called conversations of gestures, reserving special attention to the vocal phenomenon. The linguistic gesture realizes the passage from simple vocal stimulation, analogous to all other gestures, to the birth of the *vox significativa*, that is the word proper. The word is bred and fertilized by the community's response or, as Charles Peirce (1965) would put it, by the habits of conduct which are triggered by the vocal gesture (as it happens in infants).

Note: the vocal gesture is itself an exosomatic phenomenon; it resounds "out there" and, thanks to the self-affective nature of the voice (I cannot see myself, but I hear myself shouting, etc.), it rebounds on the speaker, who perceives it along with all those who are able to listen to it. In this rebound, the vocal phenomenon brings with it the common response which thereby arouses and acquires a shared meaning, internalized by all speakers. That's how I tell myself exactly and simultaneously what I say to you, which is the root of linguistic intersubjectivity and the world of mind (*Geist*, as Hegel would say). Words do not say things—this is what pre-philosophical common sense believes—how could they? How could vocal articulations reflect the structure and matter of so-called "real things"? Precisely this was, for example, the problem of Wittgenstein (1922). Words instead signal, as Peirce points out, what one is ready to do with others, that is the practical consequences triggered by language. They should not be understood as an empirical passage from theory to mere practice: practical consequences are the whole of the word's logical meaning. As Husserl observed in *Krisis* (Husserl 1959), even theory is work—it coincides with "theoretical practice".

In this brief and succinct description, the essential is that the word is the means and *medium* that transforms a response into the shared place of knowledge. This is how the work of language generates the social patrimony of knowledge, whence the nexus of the known with the unknown and the path of research. This path is made possible by the "golden" heritage (as Vico would say, see Vico 2012) of our meaningful expressions, implicit encyclopedia of knowledge, and practical wisdom. In short, language is a social machine alongside countless other social machines, from the primitive stick of the bushmen to the Geneva synchrocyclotron. We must reverse the "spiritualistic" and substantially "Cartesian" superstition that views language and culture as being produced by some kind of psychic interiority. There is nothing literally "internal" to the subject: the subject itself results from the internalization of social responses to the voice. Interiority comes down to knowing how to use the linguistic instrument acquired in millennia by the living speaker, as Aristotle would say. Similarly, the so-called inwardness of conscience, or self-consciousness, is not the fantastic product of neuronal actions. While such actions are useful starting points for a fruitful parallel reading of expressive events and linguistic phenomena, it would be a misunderstanding to regard them as their cause.

Culture is a machine, a self-propelled phenomenon and, in this sense, an automaton.[1] There is nothing improper in thinking so, no offense to the creativity of the human spirit and the sublime height of its most characteristic intellectual products: art, religion, philosophy, and science. Culture is "spiritual" because it is artificial, artfully produced (*téchne*). It consists of a living action's ongoing disembodiment and punctual re-incorporation or transcription into "artificial" bodies (disembodiment and re-incorporation are concurrent moments of the same gesture). In this regard, the advent of that marvelous machine, almost unparalleled in the course of human evolution, is emblematic: the transcription of oral communication in the letters of the alphabet. Taking shape in the graphic practice of letters-atoms, oral practice loses its original body—its expressive, gestural, mimic, and emotional body, which is canceled or suspended by being transcribed analytically in the scriptural and phonetic body of letters. This event coincides with the birth, in Greece, of the "logical mind", "historical science", and philosophy, a subject I have extensively covered elsewhere (see Sini 2009a, b, 2016). Similarly, the future of human culture has already found a very significant transcription in robotics, whose general and specific outcomes are not easily imaginable.

Culture, we said, is an automaton, because it moves by itself. Nobody decides which language to speak; in this sense, as Heidegger (1977) observed, it is language that speaks to me and through me. Nobody decides the transformations of language; they happen as any other social phenomenon, that is, by the confluence

[1] With the notion of automaton, I mean what can move by itself, by having in itself the very principle of movement.

and interweaving of innumerable and unimaginable factors related to the life of the communities and the effects of their being constantly at work. It is social work that causes unremitting changes, displacements, and innovations. By acting in invisible and profound ways, they result in unforeseen and involuntary consequences, which require a reorganization of the whole field of human experience and a correlative effort of adequate understanding and expression. This is why we do not live and think like our grandparents; and so, replacing paper with computer screens demands radical transformations in the way of promoting and understanding education in schools, to mention one consequence among others.

Human beings are the product of their "art". They are necessarily shaped and, so to speak, conditioned by it. As Foucault (1969) observed, you cannot say or think the same things at different times. Since this historical consciousness has gradually imposed itself in Western culture, investigating its "material" reasons is an important task. It is known, for example, that these new historical figures of consciousness and thought would not have arisen without the great revolution of text formation—as Ivan Illich taught us (Illich 1993)—and, on that basis, the invention of the press.

Historical and social conditioning raises perplexity in those who pose the problem of human freedom in superstitious and aporetic ways. It is not possible to elaborate on the topic here, given the length of an article; I limit myself to suggesting that freedom is not the thoroughly subjective will many imagine and presume it to be. Free is the unpredictability of what happens because what happens always involves the bottleneck of individual opportunity, in its interweaving of multiple and substantially unimaginable components and motivations. Free is, for example, the unpredictable occurrence of these words which I feel inclined to write. Having written to you, I now know that I wanted to do it, although this "will" expresses a decision that happened in me by its own (I would not say "my") freedom. My writing ego and its product are less the cause than the consequence of this impersonal freedom. At the same time, they are engendered by the cultural and social intent of participating in a collective publication of our days, and so on and so forth (all impossible and senseless things, if we think of attributing them to an ancient philosopher or medieval monk).

Social work has been underway since the emergence of human beings. This is why they are also "social": the two things, we said, are one. This work generates knowledge, namely, a technological transcription and invention of machines that fathom and, at the limit, modify and recreate our so-called "natural" environment. "Nature" is the universal presumption which has always been reinterpreted and reincorporated in the work of our historical-scientific culture (or, formerly, our mythical culture). Work translates nature into its writings as an invariably provisional image of the past from which we believe to be coming.

In the same way, it is important for us to understand that there is no "human nature" in itself: for the reasons mentioned above (the infinite proliferation of "machines", etc.), the human being is a path and a destiny, not something that "is", but something that "becomes" in an unstoppable process. Human beings are committed to becoming human and, to this end, they have at least two instruments which constantly influence and modify each other. On the one hand, there is the technical-intellectual process of scientific knowledge; on the other hand, there are the social discourses which enable us to share evaluations, choices, and goals. Words, as I mentioned above, do not say things, but establish the habits of the community; in our time, such habits increasingly take on the aspect of their global and supranational effects.

It is in this context that philosophical reflection, along with other discourses, provides its contribution. We need maps and scriptures that clarify in ever more effective ways where we come from and what, therefore, characterizes us as the beings we are. This widening of perspectives stimulates decisions capable of liberating us from the anguishes of withered and inadequate thoughts and convictions. It frees us from those old, losing habits of behavior which are often harmful in the results they obstinately set themselves to achieve. In particular, we realize that the progress of our technical power, the extraordinary results of our machines and robots, provides us with new responsibilities towards our small planet—our only real and indisputable heritage—and towards the extraordinary life that was born in it, with all its fragile creatures. Among these lives there is ours, of course. As Whitehead (1938) said, in fact, it is above all necessary to live, and then live well and live better—more peacefully, wisely, and, if possible, happily.

References

Foucault, M. (1969). *L'archéologie du savoir*. Paris: Gallimard.

Heidegger, M. (1954). *Die Frage nach der Technik, in Vorträge und Aufsätze*. Pfullingen: Neske.

Heidegger, M. (1977). *The question concerning technology and other essays*. New York/London: Harper and Row.

Husserl, E. (1959). *Die Krisis der europäischen Wissenschaften un die transzendentale Phänomenologie*. L'Aja: Nijhoff.

Illich, I. (1993). *In the vineyard of the text*. Chicago: The University of Chicago Press.

Longo, G. (2003). *Il simbionte*. Roma: Meltemi.

Mead, G. H. (1934). *Mind, self and society*. Chicago: The University of Chicago Press.

Peirce, C. S. (1965). How to make our ideas clear. In C. Hartshorne & P. Weiss (Eds.), *Collected papers*. Cambridge, MA: Harvard University Press.

Sini, C. (2009a). *Ethics of writing*. New York: Suny Press.

Sini, C. (2009b). *L'uomo, la macchina, l'automa*. Torino: Bollati Boringhieri.

Sini, C. (2016). *Inizio*. Milano: Jaca Book.

Vico, G. (2012). *La scienza nuova*. Milano: Bompiani.

Whitehead, A. N. (1938). *Modes of thought*. New York: The Macmillan Company.

Wittgenstein, L. (1922). *Logisch-philosophische Abhandlung*. London: Routledge/Kegan Paul.

Part II
Philosophy of Law

Chapter 7
Computation, Cybernetics and the Law at the Origins of Legal Informatics

Giuseppe Contissa, Francesco Godano, and Giovanni Sartor

Abstract The present contribution aims to address the ways in which legal philosophy approached and conceptualised the emergence of information technologies between the 1960s and 1970s. We will do that by examining the contributions of four thinkers, coming from different philosophical and ideological backgrounds: Vittorio Frosini, Mario Losano, Luigi Lombardi Vallauri and Renato Borruso. These authors look into the evolution of law and technology from different perspectives that are representative of different approaches to legal theory: the evolutions of idealism (Frosini), analytic philosophy combined with positivism (Losano), anti-positivism in combination with ethical/axiological inquiry (Lombardi Vallauri), as well as the more practical policy-oriented approach of legal practitioners (Borruso).

Keywords Cybernetics · Legal informatics · Computing · Artificial intelligence · Legal philosophy

7.1 Introduction: Information Technology, Cybernetics and Philosophy

In Italy the first philosophical analyses of information technology from a legal perspective go back to the end of the 1960s. At that time some legal philosophers started reflecting upon the possibilities of applying information technologies in the legal domain, envisaging what changes this would bring to the legal system, to legal

G. Contissa (✉) · F. Godano · G. Sartor
CIRSFID – Alma Mater Studiorum Università di Bologna, Bologna, Italy

European University Institute, Florence, Italy
e-mail: giuseppe.contissa@unibo.it; francesco.godano@unibo.it; giovanni.sartor@unibo.it

S. Chiodo, V. Schiaffonati (eds.), *Italian Philosophy of Technology*, Philosophy of Engineering and Technology 35,
https://doi.org/10.1007/978-3-030-54522-2_7

science, and to the very idea of a law-governed society. Such reflections preceded computerization of the law—and in that sense they are completely speculative, rather than based on any real experience—even though more or less at the same time, some Italian lawyers, in particular magistrates, started thinking about using computers for the purpose of information retrieval and even to arrange some experiments, and to develop some reflections upon them.

At that time multiple tensions divided Italian legal philosophy into opposite camps. Some thinkers were still rooted in the idealistic philosophy—which had been dominant in pre-war Italy, thanks to the work of philosophers such as Croce and Gentile; others had rejected that approach, embracing analytic philosophy and neopositivism. Some legal philosophers were rooted in natural law thinking (the view that the existence of certain norms does not depend on the will of the legislator or on any social fact, but on their intrinsic goodness or rationality), especially in the Catholic version of it, while others adopted a variety of positivistic approaches (according to which the existence of norms depends on the legislator's will, or rather on empirical social facts). In one further line of conflict at that time, those who had embraced the Marxist criticism of bourgeois law and even the idea that law and state were soon to be overcome by communist revolutions, were set in opposition to those who endorsed liberal values and a market economy.

In this complex constellation of philosophical and ideological views and conflict, some thinkers from different ideologies and perspectives focused at that time on technological evolution. The perspective of technological evolution was indeed present at the time in the social reality, as Italy was completing the transition from an agricultural to an industrial economy. It was also present in the philosophical debate, where a number of critiques of technology—in particular originating from German culture—were discussed in Italy. Among them we can mention Heidegger's view of technology as determining the way in which reality manifests itself in our era, namely as a raw material available for production and manipulation. A somewhat related perspective was also developed by the critique of the Enlightenment and of instrumental reason developed in the Frankfurt school, by authors such as Adorno and Horkheimer (and with important variations by Habermas and Marcuse). On the other hand, the connection between human nature and technology was emphasised by authors such as Arnold Gehlen, and the value of science and means-end reasoning was affirmed by Hans Albert, building upon Karl Popper's approach.

According to Losano, until 1975 Italian lawyers had a positive view of the computer, seen as a possible solution for the problems of management of the ever-growing legal documentation. After 1975 that stance changed, for many reasons, among which the 1968 movement, which indicated the computer as the essential instrument for the capitalist organization of work (Losano 1994). The trend is similar to that of other nations: concerning the adoption of computers and ICT in the legal domain, Lothar Philipps distinguishes a period of "technological optimism" (until 1975) followed by a period of "technological pessimism" (Philipps 1993, p. 13). From this moment, the interest in applications of computers in legal domain is linked with the concerns for the involved risks and the requests for legal measures to control such applications. In particular, lawyers saw the computational advancements as a risk for their intellectual expertise and their social role (Lombardi Vallauri 2014, p. 17).

We cannot here consider the general philosophical debate on technology unfolding in Italy at the time. We shall rather focus on the four main legal thinkers who addressed the emergent phenomenon of the application of technology to the law: Vittorio Frosini, Mario Losano, Luigi Lombardi Vallauri and Renato Borruso. These authors looked into the still-to-come hybridisation of law and technology from different perspectives that are to some extent connected to their different approaches to legal theory: the evolutions of idealism (Frosini), analytic philosophy combined with legal positivism (Losano), a critique of legal positivism, in combination with ethical/axiological/theological ideas (Lombardi Vallauri), as well as the more concrete, policy-oriented, and revolutionary approach of a practitioner (Borruso).

Their views have a historical significance, but they are still relevant today. It is surprising that the issues that these authors raise at a time, between the end of the 1960s and the beginning of the 1970s, when computer still had no real impact on the law, have not been discussed in later times, when ICT acquired a much more important role in legal activities—in courts, administration, legal practice. On the one hand, this is due to the fact that mainstream applications of IT in the legal domain—from digitalisation of legal sources to the automation of registries and administrative activities—no longer attract attention, as the accepted common-sense environment; on the other hand, such applications have not really engendered deep change in the core of legal activities. However, the situation is changing nowadays, with the massive introduction of AI in the legal domain, for purposes such as the analysis of masses of legal data, the prediction of legal decision, the automation of the application of the law in large domains. Technology is now challenging the functioning of the law and the way in which we understand it, as these authors anticipated a long time ago. Therefore, their analyses may still be relevant to imagine and design the future of law and technology.

7.2 Vittorio Frosini: A Legal Reflection on Cybernetics

7.2.1 Cybernetics and the Law

In 1968, Italian legal philosopher Vittorio Frosini delivered what is probably the first extensive account of the relatively novel subject matter of computing technologies in its relations with the law. He reflected on such technologies adopting the approach of "cybernetics", a discipline which had a great potential in the transformation of the means of understanding and governing different social systems—including the legal system. Significantly, the book was entitled *Cybernetics, Law and Society*, and it is worth noting that this first work on IT and law was published in a series dedicated to philosophical and sociological issues in the law.[1]

---

[1]The series was issued by the *Edizioni di Comunità* publishing house, and was named *Law and Modern Culture* (*Diritto e cultura moderna*). It was directed by two prominent scholars, legal sociologist Renato Treves (1907–1992) and legal philosopher Uberto Scarpelli (1924–1993).

The study of information processes and their technical applications flourished in the midst of the twentieth century within the wider scientific perspective of cybernetics. Cybernetics was conceived as the study of "communication and control systems" pertaining both to physical dynamics and to living organisms, to the natural as well as artificial domain.[2] Cybernetics meant to use the theory and technology of information in order to understand how information acts in the "control" of the systems. Furthermore, as Wiener strongly supported, the investigation could apply to both mechanical and biological systems, and establish a comparison between the two.

Wiener was probably the first to identify the relationship between cybernetics and the law. He believed that legal problems are by their very nature communication and cybernetic problems; that is, they relate to the regulated and reproducible govern of certain critical situations (Wiener 1950). This sentence clearly captures the attraction of the law to the cybernetic framework, and explains why information technologies have revolutionized the legal domain since the beginning, forcing lawyers to rethink its philosophical and theorical background as well as its tools and applications.

Frosini noted that the distinctive feature of computers, among other machines, is that they use electricity to produce information. In this context, cybernetics is not concerned with perfecting automated mechanisms, or robots, as mechanisms capable of a more complex use of energy, but it is concerned with our knowledge of control and information systems, common to animated beings and machines, in order to increase our intellectual power over the world (Frosini 1968, p. 93).

7.2.2 The Cybernetic Ideology

In the introduction of his work, Frosini states that his aim is to analyse some of the problems caused by the "technological revolution" in the fields of the law, of society *and of human consciousness* itself (Frosini 1968, p. 9). He recalls Merleau-Ponty (1964, p. 12) who introduced the term "cybernetic ideology" to define the spiritual condition of a technologically advanced society, where consciousness is depleted by the artificial, merely "operational", automated reason of the machines. Against such turn in the development of mankind he takes a position of alarm, protest and distrust. According to Frosini, such position was very indicative of a state of mind, still very widespread among intellectuals of humanistic tradition of his times, against the new cybernetic civilization. Another example of such stance is the approach of Marcuse towards automation in society: "The machine process (as social process) requires obedience to a system of anonymous powers—total secularization and the

[2] In short, the main features of cybernetics can be described as: (1) the study of information processing in order to understand the goal-oriented behaviour of systems; (2) the application of this method to both physical systems and biological systems; and consequently (3) its application to both animal-human realm and to artificial objects (machines). See Wiener (1948).

destruction of values and institutions whose de-sanctification has just begun" (Marcuse 1964, p. 50). According to Marcuse, this process would determine a flattening of human consciousness, or rather its reduction to a single dimension, the external one or of social relations, with the loss of interiority; in short, the human being would become, as Merleau-Ponty suggested, a *manipulandum*.

Against these positions, Frosini does not assume the term "ideology" as a negative qualification. Instead, he considers the "cybernetic ideology" as expressive of a new attitude which the human mind is obliged to assume in the face of the transformation of ways of life, determined by the advent of the second industrial revolution, and which is broken down and manifested in the most diverse fields of human activity (Frosini 1968, p. 69). The progressive adoption of computers in productive systems, and more broadly in every social process, is the sign of an economical, technological and social transformation marking the advent of a *new dimension* of human activity, rather than its reduction to a single dimension, as claimed by Marcuse. In other words, the age of information technology shall instead—just *through* the automation of certain activities—enhance human consciousness, enrich its view of the world and its critical ability. The widespread adoption of computers can be considered the symbol of such revolution, but its wider implications can be understood only analysing the phenomenon of automation of production processes, that is the "cybernation", adopting the neologism coined by the sociologist Micheal (1962). Recalling a definition by Valerio Tonini, in this context Frosini sees cybernetics—considered from a unitary point of view and inclusive of its social interest—as the new science of a civilization of machines and automation, in which science and technology are considered the essential factors that will create the future condition of mankind (Frosini 1968, p. 77).

7.2.3 Cybernetics, the Law and the Lawyer

Legal philosopher Sergio Cotta considered the jurist a "technologist" of a particular kind, "the expert of social life". Our technologically-driven society, he claimed, is not organized around the political or military power anymore, but around the productive apparatus built in union with science and technology (Cotta 1966, p. 286). Drawing from this thought, Frosini affirms that the extraordinary proliferation of the artificial tends to make the human being *a second-degree agent*, who animates and directs acting machines. The philosopher believes it appropriate that in Italy, just like in the US and other advanced countries, there should be an adequate awareness of the new reality on the part of the cultural avant-gardes, and that it is up to the jurist, because of the importance of the social function he has traditionally performed in our system of civil coexistence, to assume that responsible attitude of moral guidance, which elsewhere is attributed to the "social scientist".

Frosini observes the expansion of the methods of automation and the instruments for their command and control. On the other hand, the very boundaries of human thought and action are expanded through the adoption and use of the computer, which becomes a *miroir de l'homme*, since its compositional structure tends to become a mirror image of the human mind. In this context, according to Frosini, cybernetics offers the possibility of modifying the methodological foundation of the human sciences, making use of the technological experience derived from cybernetic research: by shifting scientific interest to the problem of "communication", cybernetics has created, so to speak, the foundation of social knowledge on the ground of artificiality, that is, it has enabled the transformation of cultural paradigms into "communication and control paradigms".

The new cognitive method, or rather the new interpretative principle of the processes of knowledge brought by cybernetics, shall then be extended to the social field including the legal domain, where Frosini supports the adoption of computers, and of the new methods of command and control of information. Against those critics fearing the introduction of an "artificial law" characterized by the widespread adoption of computers to support legal activities and tasks, Frosini replies that law can be said "artificial" only in an arbitrary manner, that is, in so far as it is opposed to "natural" law, but not in so far as it means the use of technology for the law: the artificiality recognized (or feared) as a characteristic of law in the age of automation is, after all, nothing more than a metaphor used to indicate its *logical-formal moment*, which is itself constitutive of legal experience just like the ethical moment. According to Frosini, the application of technology, understood as a process of operational rationalization, constitute the new shape of a world, in which humanity is taking on a new spiritual form, so that the technological mentality becomes its second, renewed nature. Therefore, those new perspectives of thought and experience, which the general scientific progress also presents to lawyers, must be accessed and analysed (Frosini 1968, p. 39).

In the legal domain, the problem of adoption of computers and of the cybernetic methodology has been posed, but the approach does not appear complete since the gap and tension between the legal factual reality, its changing morphology of operational structures and its formalisation in computer models have not been fully considered. Thus, cybernetic ideology, in the positive meaning given by Frosini, represents a need, which is increasingly felt, for a greater systematization, i.e. rational and operational coherence, between the rules of behaviour in civil coexistence: the advent of a social system, in which the structures of human relations are recomposed according to rational lines of force, and whose representative ideology is intended to promote the exercise, progress, innovative initiative of human reason.

According to Frosini, the question of applicability of cybernetics to judicial activities or the application of the law is not the question of determining whether "machine-made" law is logically admissible, or whether it is appropriate to set up a "spitting machine" in the courtrooms; rather, it is a question of observing what kind of impact may results from the deployment of technologies and innovations in the field of cybernetics to the organization of justice.

Frosini claims that the advent of computers in legal retrieval activities will result in a profound reform and restructuring of the traditional and rigid hierarchical system of classification of legal rules, which should be rearranged on the basis of an understanding of their organic and rational correlations. If this result would be obtained, i.e. to rationalize the system of juridical production, then the computer would prove to be a device of universal utility, an "electronic razor of Occam" on the basis of the old principle that *entia non sunt multiplicanda sine necessitate*. According to Frosini, another important link between cybernetics and law is the area of "logic of probability", that is, the application of classical statistical method to the analysis of behaviours emerging from the law, consisting in a quantitative assessment of legislative effectiveness, taking into consideration the analysis of quantity and frequency of the practical consequences of a rule or of a legal relationship. Thanks to computers, it would become possible not only to investigate the past in order to better understand the existing situation, but also to formulate statistical forecasts for the future, which have a very high degree of approximation (provided, of course, that certain environmental conditions remain constant).

Focusing on juridical automation, Frosini is interested in the possibilities to automate juridical tasks. However, he is also concerned with the correlated issues: a "machine-made justice", he claims, runs the risk of entailing new fetishisms and increasing the conceptual rigidity of the jurists; it might favour the escape from personal responsibility, and threaten to reduce personal liberty, undermining its freedom of judgment (Frosini 1968, p. 25). In particular, Frosini highlights that the computer is unable to model the dialectical process, which is the core of legal reasoning. At most, the computer would be able to analyse a computable representation of it, based on formal logic. The language of mathematics and propositional logic is based on pre-determined schemas, which have to do, respectively, with the relationships between numbers, and between propositions, but not with their internal nature or meaning (Frosini 1968, p. 103). The legal language, instead, consists in a symbolization of needs and operative structures of human practice in social life, being the point of passage between the changing and disordered reality of the human praxis, and the structure of the legal order, which reflects and orders it in a *logos*, that is, in a coherent and finalized discourse. It is therefore necessary that the attempt to adapt practice to a reasonable order, such as that of law, also corresponds to the capacity of law to adapt to the living forms of human action, that is, the legal language needs to maintain a considerable degree of elasticity in order to be used in its practices. In conclusion, according to Frosini, it is not possible to reduce the law to the sole dimension of rationality, because it is not possible to reduce social life itself to such dimension.

7.2.4 Artificial Consciousness and the Human Being/Machine Relationship

According to Frosini, the development of computers, and, in the future, of artificial intelligent systems, is triggering a reversal of perspective, with regard to the traditional problem of the relationship between human being and machine. In the past, the main research topic was to provide a mechanized representation of the human being, seen as a complex mechanism of which the highest manifestation was consciousness; instead, today the problem is providing a humanized representation of the machine, that is, establishing to what extent the consciousness of the human inventor may be transferred to the machine she/he invented.

In other words, the new question is whether a machine may be aware of its actions, that is, whether consciousness can be generated as an artificial product. Frosini observes that the introduction of the Darwinian theory of evolution, despite depriving mankind of its "theological" speciality, yet confirmed that the human being represented the highest step of natural evolution. However, the technological revolution puts into question the superiority assured to him by the "consciousness" (last dominant step of the evolutionary scale) on the rest of the world, animated and inanimate. But, if it is possible to create machines equipped with consciousness, it is possible to go beyond the creator herself/himself: echoing Pico delle Mirandola's link between animal and angel, the human being would thus become—as Butler (1863) had suggested—a link between the world of nature and the world of technology.

The human being is at risk of losing her/his cognitive superiority: the inventor, constructing her/his machine, wants and has to limit herself/himself, so as to allow the machine an effective freedom of choice (autonomy). Moreover, in a context where machines are autonomous and substitute humans, Frosini sees the risk that their errors of prediction may be enormously multiplied in their effects, with irreparable consequences: a machine equipped with its own purpose will pursue it without taking into account any practical meaning, which must be present in every purpose of the action, and which only the human mind can establish, evaluate and correct. The solution, according to Frosini, is not to focus on the computer as a substitute for humans, but rather to aim at achieving (as also suggested by Wiener 1950) a sort of intellectual symbiosis between human being and machine, based on the distinction of tasks and cooperation of functions between the two components (Frosini 1968, p. 101).

The computer does not *reproduce* the human thought, but rather *mimics* it, Frosini argues: it is a *simia hominis* that reproduces human thought's movements, rhythms, procedures, by means of algorithmic procedures (Frosini 1968, p. 104). However, there are non-algorithmic mathematical problems, whose solution (complete or partial) can only be obtained in a non-algorithmic way, that is, by resorting to an intuitive mental procedure, and not an analytic one: this establishes a gap that cannot be overcome between the computer and the human being.

Frosini then asks to what extent the knowledge of computers helps us shed light on the knowledge of human thinking processes. The consciousness of the machine, according to him, cannot have intentionality in the full, human meaning of the term, which implies a purpose, originated by a moral need. Despite such limit, Frosini claims that it is still legitimate to speak of a "mechanical" consciousness, meaning those mechanisms of behaviour and perception that can be explained in cybernetic terms, and that make up a conscious substructure, which he calls the "artificial" consciousness. Looking at the perspective of symbiosis between human being and machine, he foresees the possibility of a transfer, or even perhaps an interchangeability, of mechanical consciousness between human being and machine, without the replacement an artificial consciousness with a "natural" one (Frosini 1968, p. 107–109).

7.2.5 The Human Condition in the Age of Cybernetics

Frosini observes that in modern society the human being continually passes between two forms of consciousness, the *internal consciousness*, that is, the awareness of being herself/himself and acting, and the *external consciousness*, that is, the awareness to play the part which has been assigned to her/him, according to legal norms, social norms, and institutions at the basis of civil coexistence. For example, the judge, when reconstructing the case brought before her/him, proceeds in the dimension of the external consciousness, which is that of the "man of law", in accordance with the rules of procedure. However, at the same time a different consciousness may well arise: the judge's awareness that her/his actions are correct, indeed scrupulous, because she/he is invested with a moral burden of responsibility, that is, her/his internal consciousness.

A strong tension exists, according to Frosini, between the feeling of possession of one's own original, natural humanity and the feeling of belonging to a condition of artificial life, that is, dominated by the forms and needs of the industrial society: a tension where the human being feels the danger of such a strengthening of her/his *external* consciousness, increasingly modelled on the image of an artificial world, that would compromise the vitality of her/his *internal* consciousness. However, the author affirms, this artificial world is not an inhumane world. On the contrary, it is the creation of human consciousness, desired and made by mankind, to whom it is therefore fully homogeneous, much more than the natural world itself; the world of technology is in fact the true world of mankind, the world of society in which the operative reason unfolds. The contrast, which undoubtedly exists, is therefore the contrast of the human being with herself/himself, which has taken the place of the ancient quarrel between "natural" and "spiritual" human being: today's human being, according to Frosini, is indeed herself/himself an artificial human being, because our natural side has been constantly fed by our capacity to build artefacts which enabled us to protect and improve our "natural selves":[3] what makes history human, and makes the human being a historical being, is precisely the capacity that the human being has to "manipulate" herself/himself, that is, to transform herself/himself by creating the "artificial" world—a world that includes nature-altering science and technology, art itself that is artistic because it is artificial, as well as the laws that govern human society.

[3] We can note how this need of protecting and improving is the same drive of religious and philosophical thought.

In this context, according to Frosini, the cybernetic programme—focused on the problems of information and goal-related control—supports us in recognizing and maintaining the constant relationship of correspondence and conversion of the natural into the artificial: we can understand the functioning of the eye thanks to the progress made in the field of optics with the invention of lenses, precisely because of that process of mediation, which allows us to return to look at the eye passing by the experience with technological devices. In the same way the computer, which can read and understand digital information, performs the operation of a new kind of eye, an artificial and spiritual eye, built by humans for humans. Frosini here recalls the ideas of his master Ugo Spirito, who observed that the machine itself can only be conceived as a development of human nature, with all its attributes; as such the machine can be said to be "artificial" in an arbitrary manner only, that is, by *arbitrarily claiming the end of the process of nature at a certain point of its technical evolution* (Frosini 1968, p. 39, 87).

7.3 Mario Losano: Legal Technologies and the Legal System

7.3.1 Information Technology and Legal Theory

Together with Vittorio Frosini, Mario Giuseppe Losano was a pioneer in the studies of information technology and law. Legal philosopher and legal theorist, Losano applied the analytic method to the reflection upon legal systems and fundamental structures of the law. Hence, his interest in the emerging cybernetic domain was mainly concerned with its impact on the theoretical framework of the law as a social system. He was among the first scholars to survey the studies in computation and law of his time, and to give an account of the projects related to legal informatization which were being established in Europe, in the USSR and in the United States.[4] The survey was part of his 1969 book entitled *Legal Cybernetics* (*Giuscibernetica*),[5] the first systematic work specifically dedicated to models and applications of cybernetics in the legal domain.

Losano believed that the cybernetic approach to the law had a potentially strong impact on legal-theoretical issues, since it allows the lawyer to think of rules, of the relations between the legal actors, of the boundaries of law with other human "systems" of activities as problems of system govern through information. The author defined "systematics" the activity of seeking order in the legal system, i.e. of arranging the law according to the model offered by a certain legal theory, or under certain paramount principles of justice, in order to assess the justice, consistency and com-

[4] A considerable part of his main book on information technology consists in fact in a detailed description of the existing projects classified by country (Losano 1994, p. 47).

[5] The complete title is *Giuscibernetica. Macchine e modelli cibernetici nel diritto* (*Legal Cybernetics. Machines and Models of Cybernetics in the Law*), Torino: Einaudi, 1969.

pleteness of the system itself (Losano 1994, p. 35–36, 87).[6] This activity was neglected, he claimed, by the uses of computation technologies to the legal field at the time, basically related to the "jurimetrical approach" of Anglo-American origin, which we shall see in what follows. Consisting mainly in a quantitative analysis of legal data, those uses—while suiting a common-law based system—amounted, according to Losano, to an "insufficient employment" of the computers' potential, especially when transplanted on European soil.

7.3.2 *Jurimetrics and Its Cybernetic Evolution*

Losano observes that the first applications of electronic computation to the law were taken up by jurimetrics, a discipline which analysed legal matters through mathematical and experimental methods. Jurimetrics was born in the United States in the 1940s as an attempt to give a more rigorous, measurable character to the law. Its founder, Lee Loevinger, considered jurimetrics an effort to utilize the methods of science in the field of law. He opposed this new field to the speculative, philosophical realm of jurisprudence, whose object is the quest for ultimate answers concerning essence, ends and values of the law, which cannot be considered scientifically. On the other hand, jurimetrics "investigates methods of inquiry", and "is concerned with such matters as the quantitative analysis of judicial behavior, the application of communication and information theory to legal expression, the use of mathematical logic in law, the retrieval of legal data by electronic and mechanical means, and the formulation of a calculus of legal predictability" (Loevinger 1963, p. 8).

Discussing the jurimetrical approach, Losano (1994, p. 94) emphasized that Loevinger employs a concept of science which only includes what is mathematical, experimental, quantifiable. This idea, while echoing the debates of the Enlightenment and the conceptual legal systems of the Pandectistic school, does not fit the current European cultural milieu adequately. Recalling the prominent Italian legal theorist Bobbio (1955), Losano affirmed that legal science *is not* mathematics in its essence (he uses the term "mathematics" to simplify the concept of quantifiable scientific method); however, mathematics can *serve* legal science.[7]

The scientific perspective of jurimetrics is reflected in its object. The main research and projects of jurimetrics consisted in electronic data processing and its

[6] Here we can disregard Losano's distinction between "systematics" and "potential practical activity" (Losano 1994, p. 35–39), which is not needed for our purposes.

[7] Loevinger's quantitative approach reflected the idea of analytical philosophy that reality—and therein the legal domain—can be reduced to language. On the point, Frosini (1968, p. 59) highlighted that the legal phenomenon appears irreducible to the linguistic profile alone, in which it was forced by the methodology of the analytical school, and it may instead find an interpretative formula in the cybernetic models. Frosini noted that cybernetic models are based on schemes of energy transmission, and, in the same way, law is composed of messages, i.e. normative expressions, being essentially a complex of structures of human action in social life, which are symbolized and made communicable by means of a code of signs.

use to classify and analyse case-law data and predict the behaviour of tribunals statistically. This work includes the basis of any legal computing endeavour, and is therefore essential in the cybernetic framework: however, Losano highlighted that it offered only a partial view of the legal mechanism when transferred in Europe. Courts' data analysis is in fact essential in the Anglo-American common law system, whose "legality" is essentially constituted of the tribunals' pronouncements. On the other hand, the production of norms in the civil law systems of continental Europe results from a combination of written rules, case law and, to some extent, scholarly works, and their "computability" required a more complex mechanism (Losano 1994, p. 44).

Therefore, the author (Losano 1994, p. 108), as well as other scholars (e.g. Lombardi Vallauri 2014), concluded that the use of IT as quantitative analysis had to be extended to other "factors" of these legal dynamics. Jurimetric research and technologies were not to disregard: on the other hand, as stated above, their domain is essential in the cybernetic programme. Accordingly, the jurimetric approach—which was later referred to as "documentary informatics", mainly concerned with data processing, storage and retrieval, had to be included in a wider set of tools, which shall consider:

1. computable law as a system (legal theory), in its internal workings and in its connection with other systems, political (particularly in the development of democracy), economic, and so on;
2. the use of IT to support and enhance the core intellectual activity of the lawyer: legal interpretation and decision-making.

The latter is what Losano categorized as "legal cybernetics" as opposed to jurimetrics, and was later known as "meta-documetary informatics". The perspective captured the deployment in the law of what many considered to be the potential "wide research programme" of cybernetics (see for instance Somenzi 1993). Through the "artificial" means of computation, in fact, this wide perspective was meant to explore—and to refine—the "natural" mechanisms of communication and govern of the law. We may say that, despite originating in just that cultural milieu, the full unfolding of such programme in the legal domain was not encouraged by the Anglo-American juridical structures. On the other hand, continental European civil law seemed to offer a more fertile terrain to further uses of computable model of the law. Such models could capture not only the internal structure of legal systems, but also the communication and feedback processes concerning the interaction between the legal systems and its social environment. Even though legal processes can be highly automated, human intervention is still decisive to "govern" the processes of communication and action affecting the legal system and its relation to society (Losano 1994, p. 111).

7.3.3 *Cybernetic Models of the Legal System*

According to Losano, the application of information processing to "systematics" constituted a clear fulfilment of the cybernetic programme (Losano 1994, p. 106–113). The main limit of jurimetrics, he observed, was in fact that it only allowed to work on single rules or group of rules; cybernetics, on the other hand, meant tackling the law as a comprehensive system of rules.

Such a perspective opened different directions for research. First, it enabled a more accurate analysis of the legal system as a whole and as a sub-system of the social body. Losano observed that the relations between the legal systems and society is shaped by the dominant philosophical and political approaches. For example, in capitalist and socialist societies, the State and other "normative agents" adopt different models of regulation, social groups (industry, professional lobbies, politics, and so on) influence the law to different extent and with different modalities. Cybernetic models would give a decisive support in examining and governing these dynamics. Losano showed how these models were used in the socialist countries of his time (namely the USSR), which considered the State and its legal apparatus as a means of governing economy. Cybernetics, he claimed, was developed here as a "science of control" of society through the law (Losano 1994, p. 114).

Another use of cybernetics in this domain, according to Losano, concerned the legal system in its internal working, i.e. the interaction between sets of norms, legal agents and institutions. Cybernetic models would here enhance the study and the practices of the law as a self-organizing system, in its quest for equilibrium in a dynamic evolution and in the relations with other parts of society. The author recalls Knapp's theory of regulation and perturbation (Knapp 1964), and Ballweg's studies of the legal system as a feedback system (Ballweg 1965).

Finally, the "systematic" side of the law would highly profit from a "jurimetrical", quantitative analysis extended to the design of legal systems as a consistent set of rules, and to the subsequent task of tackling the issues related to contradictions between rules and to legal gaps. This aspect would be of particular importance today, Losano adds, in view of the growing need of rule harmonization of national law in the European framework (Losano 1994, p. 87–89).

7.4 Luigi Lombardi: Ethics and Axiology of Legal Computation

7.4.1 *Philosophical Reduction of Technology and Cybernetics*

Luigi Lombardi Vallauri is one of the most eclectic figures in twentieth century's Italian legal thinking. He made significant contributions in the fields of philosophy, legal theory, legal history, bio-ethics, religion and forms of spirituality, animal rights, and more. In 1969, he wrote an article on Frosini's trailblazing book on cybernetics and law, and during the 1970s he dedicated himself to the new IT studies, holding the post of director of the recently founded Institute of Legal Documentation (Istituto per la Documentazione Giuridica, IDG)[8] of the Italian Research Council.

In his book specifically dedicated to reductionism as a paradigm of the modern culture, he includes "information" in the "reductionist formula", together with matter and energy (Lombardi Vallauri 2002, p. 15). Moreover, he abides by the so-called "algorithmic hypothesis of the mind", the possibility of representing the operative functions of the mind as algorithms. He considers this reductionist approach valid, and "crucial" for contemporary science: however, he maintains a "dual", or "heterogeneous" position on the ontological level (Lombardi Vallauri 2002, p. 36). Matter, energy and information are insufficient to capture the ultimate meaning of human nature, subjectivity, agency or even senses; algorithms can explain the mind's communication and control process, but they do not constitute the mind ontologically (Lombardi Vallauri 2002, p. 57).

7.4.2 Cybernetics, Law and Politics

In his comment of Frosini's work, Lombardi observed that the applications of cybernetics to the law were bound to have a strong impact on the theory of law and legal scholarship. However, the legal-theoretical aspects cannot be separated from the cybernetic impingement on the more general "theory of mankind and human thought". The two domains were presented as closely interrelated. Lombardi saw the extensive automation of the law as desirable and technologically possible. However, he claimed, such automation required the automation of human experience as a whole: and this implied endowing the algorithm with a strong philosophical character. Following his "dualist" position on the reductionist character of cybernetics, he argued that the "algorithmic hypothesis of the mind" was to be fully explored and followed through in all its potentiality, for the automated control of legal issues to be accomplished (Lombardi Vallauri and Trautteur 1969, p. 428). The lawyer as a lawyer, he wrote, is no more, nor less "automa" than the human being as such (Lombardi Vallauri 2014, p. 14).

The strict dependence of the legal domain from the integral system of human consciousness is reflected in the relationship between law and politics. Lombardi considered legal science as embedding a necessary "political moment". Therefore, the possibility of automatizing the law depended, he argued, from the possibility of automatizing the political domain (as a part of human experience as a whole). Moving from a position of "free law", namely the creativity and political nature of

[8] The IDG was one of the first institutions dedicated to automatic processing and storage of legal documents (laws, courts' decisions, administrative regulations, etc.) in Italy. In 2001 it was replaced by the Institute of Theory and Technique of Legal Information (Istituto di Teoria e Tecniche dell'Informazione Giuridica, ITTIG). See www.ittig.cnr.it.

legal decision-making, he deemed it of the greatest importance that cybernetics—and, later, IT—was employed as a tool for the development of democracy. Legal computing had a strong potential, he claimed, to reduce the "legal-technical" power of jurists, and consequently their corporative power. As such, the development of cybernetics should tackle the role of the legislator, which constitutes the chain of transmission between politics and law in civil law systems.

To this end, Lombardi affirmed, in line with the abovementioned view of Losano, that automatic legal processes had to move from their "jurimetrical" dimension of data process, storage and retrieval (documentary informatics), onto a meta-documetary informatics that would take on the turf of legal science (and politics): reasoning, counselling, decision-making (Lombardi Vallauri 1975, 1982). His vision was that of an "integral legal expert system" (Lombardi Vallauri 1995): as we shall see, this perspective translated into several projects promoted by Lombardi in the field of legal IT.

7.4.3 Automated Legal Activities

The first and foremost field of application of IT to the law was what we called "documentary informatics". Automatic information processing has a strong, intuitive capacity of easing organizational matters involving data management. For this reason, computers were employed in the legal domain for organizing, storing and retrieving data related to the main areas of the law (legislation, adjudication, administration, etc.) both in public and in private organizations. This entailed an enormous transformation in the work of legal activities, which enormously simplified the tasks related to research and analysis. However, as mentioned above, in Italy and other European systems the need was soon felt to move to a "meta-documetary informatics", whereby the computer could impinge directly on the core functions of the legal system and of the law.

Lombardi was a fierce promoter of these functions of computation in the law. In the years 1973–1977, as director of the Institute for Legal Documentation, Lombardi concentrated on the planning and organization of legal documentation systems for the Institute. Drawing on his ideas of IT and cybernetics as key factors of a full democratization of advanced countries, he strongly advocated the necessary primacy of meta-documentary informatics, which should serve as a means to render technology a tool to help citizens access public services and participate to democratic life.

Putting his thoughts into practice, Lombardi was a pioneer of a series of projects in IT and law. In 1975 he proposed a computer-based free legal counselling system for citizens, who could forward legal questions through a network of centralized elaborators, directly to lawyers hired by the Institute for Legal Documentation (Lombardi Vallauri 1975). A little earlier, in 1974, he had launched a prototype of the so-called "Automa giuridico infortunistico", arguably the first automated expert system in Italy. The prototype could provide legal advice, in the field of insurance law, on a given case submitted to it, on the basis of a legal framework and a corresponding reasoning tool. The project, however, was not further developed.

Following his view of the political-evaluative nature of legal decisions, as opposed to the logical-formal theories of law, a few years later Lombardi proposed a series of algorithms that could operate an evaluating decision in a legal case—under a paradigm of axiological non-cognitivism ultimately relying on the so-called "Huber algorithm," [9] according to which, in absence of clear indications, one should decide as they would if they were legislators (Lombardi Vallauri 1982).

As a summit of his speculative path, in 1993 Lombardi exposed the theoretical framework for the ambitious (and maybe far too visionary) project of a "Global Legal Expert System" (Sistema Giuridico Esperto Globale, SEGG). The project was meant to simulate the whole process related to the life of legal norms, from the arrangement of a philosophical-theoretical framework of a legal domain to the political-legislative activity down to the application of the single rules and the resolution of disputes thereof (Lombardi Vallauri 1995). Echoing the holistic feature of cybernetics, Lombardi's overall ambition seemed to be the conversion of all "legal-technical" expertise into the field of algorithm crafting and governing. The technological state of the art at the time—together with the plausible resistance of the "legal industry"—was doubtlessly the main restraint to his imagination.

7.5 Renato Borruso: Computers for a New Rule of Law

7.5.1 Language, Law and Information Technology

Renato Borruso was a judge of the Italian court of Cassation. He was in the team that developed the Italgiure system. In his work of 2004, *L'informatica del diritto*, he focuses on the relationship between language, law and information technology. According to the author, such relationship is the most important topic of informatics applied to the law, and it is in turn reflected in the relationship between the law, the judge and computers, finding its focal point in the interpretation of the law.

Borruso observes that the law often lacks unambiguity, analyticity, completeness, so that even jurists often find it difficult to interpret and coordinate norms. During the legislative process, the need to find an agreement based on compromise often pushes the legislator to formulate laws in an evasive or elastic way, because otherwise agreement would not be reached. Such a way of legislating increases more and more—and to a certain extent pathologically—the importance of the jurisprudential activity of legal interpretation, giving judges enormous power: through the interpretation of the law it is possible to integrate or reduce or in any case modify and, in certain borderline cases, overturn the scope of a rule, to the point that the intervention of jurisprudence may be considered creative, in the fullest sense of the term (Borruso et al. 2004, p. 293).

[9] Eugen Huber (1849–1923) was a Swiss lawyer, main author of the 1907 Swiss Civil Code. Art. 1 of the Code states this clear-cut principle of interpretation for the judge, which was meant as a "responsible constraint" to make a decision in an unclear situation.

Borruso notes that the jurisprudential activity of interpreting the law has always been seen with extreme mistrust by the legislator, and with good reasons, since interpretation can sometimes lead to corrections and/or additions to the text of the law so significative as to alter its original connotations and even, sometimes, to completely frustrate the will of the legislator. In these cases, the law depends on the individual features of the judge who is interpreting it, and therefore unpredictable in its applications. Moreover, the interpretation of a law is usually carried out not on the basis of those precise and predetermined, unequivocal and complete rules which in mathematics (as well as in informatics) are called "algorithms", but on the basis of what Borruso calls an *esprit de finesse*, so that interpretation consist not in the application of a science, but rather in the application of an art, in the performance of a creative and ethical activity (Borruso et al. 2004, p. 320).

According to Borruso, while the Italian lawyers of Enlightenment, in imposing the Philangerian conception of the law at the foundation of the tripartition of the fundamental powers of the State, conceived the law in a way similar to what now it would be called an algorithm, in the following times the model of the law moved further and further away from that of the algorithm. This fact raises a problem, Borruso claims, with the tripartition of powers (conceived during the Enlightenment but still today a fundamental pillar of Italian Constitution), which is still based on a (now) flawed assumption of the law as an algorithm.

7.5.2 Information Technology to Solve the Problem of Legal Interpretation

In this context, Borruso sees the possibility of resolving the problem of interpreting the law using information technology. In particular, he considers five main ideas that would help understand the intimate philosophical and practical connection between law, jurisprudence and computers: (1) what matters most in computers is not the machine (so-called hardware), but the written message (the programme); (2) the computer is the first and only machine that the humans can drive with written or oral words; (3) the computer has the correlated ability to read what the programmer writes for it; (4) the computer has the ability to apply deductive reasoning (syllogisms) to make an assessment on specific cases, according to the set of rules that in computer science are called "algorithms", and are incorporated into programmes. Therefore, the computer may be defined as a "*universal algorithmic automaton*" in the sense that, when a certain activity is carried out according to rules that can be expressed in a clear, concise, unequivocal, complete form (that is, in the language of the algorithm), it is always possible to make the computer carry out such activity (replacing or supporting the humans). Borruso summarises this idea in the motto: *ubi regula, ibi computer* (Borruso et al. 2004, p. 333); (5) the computer does not have, however, the ability to interpret the programme, if not taking all its instructions literally. It is therefore the responsibility of the humans, when interacting with the computer, to express the instructions so clearly, so precisely, so comprehensively that they are directly applicable without the need for interpretation.

Borruso then considers the question which, in his opinion, is at the heart of all legal information technology, in a perspective similar to Lombardi's approach (*supra*, Sect. 7.4.2). If the bypassing of interpretation is possible for the humans in relation to the computer, why should it not be possible for the legislator when enacting laws? Why not induce the legislator to formalize her/his statements with the same rigorous approach, with the same methods that the programmer uses when instructing the computer? The solution, in the author's view, is to induce the legislator to draft laws no longer in natural language, but in a *formalised* language, based on propositional logic and the use of Boolean operators, so as to reduce as much as possible the ambiguities and incompleteness of legal texts and, consequently, the interventions of jurisprudence, so as to restore to the principle of the tripartition of the fundamental powers of the State. In this way, not only would the litigation be considerably reduced, but also the legislator, confronting herself/himself with the need of expressing statements in a rigorous and unambiguous way, would have the opportunity to better reflect on her/his own will.

This approach, continues Borruso, would also facilitate the automated application of the law, supported by computers, through the development of "legal expert systems", i.e. systems enabling the computer to apply a set of (abstract) rules of law to a concrete case. They are developed by adopting the process of formalization or normalization of legal knowledge (Allen 1957), coupled with computable models of legal reasoning provided by formal logic. According to Borruso, they would be a valuable tool to bring law closer to citizens, allowing for the maximum conceivable simplification in the application of the law, summarized by Borruso with the motto: *da mihi factum, dabo tibi ius* (Borruso et al. 2004, p. 343).

With the development of legal expert systems, legal informatics would become a training ground for the refinement of legal thought, where lawyers practice to weigh the value and relationship of words. In those cases, where the automated application of the law would still be impossible, that would be caused by a wording of the law that is ambiguous, imprecise, incomplete, or, in computer science terms, that is not an "algorithm": discovering the non-algorithmability of current laws would still be very useful to the jurist because it would stimulate and accustom her/him to a critical analysis of the legal text.

Borruso claims that the final goal of the computerization of law shall be the realization of the "law-software"; the legislator should formulate the law in the form of an application software, as if it were a legal expert system, thus not only eliminating any uncertainty of interpretation, but also enabling a direct interaction between the legislator and the recipient of the law (the citizen) bypassing, with regard to the interpretation of the law, not only the lawyer or the public official, but the judge herself/himself. In this way, Borruso concludes, we would move from the age of *vigilantibus succurrunt jura*, to the era of *da mihi factum, dabo tibi jus*. In other words, from the era of the restricted knowledge of law by a privileged few, to the era of true democracy of law.

7.6 Conclusions

In these pages we have analysed the way in which four leading legal scholars have interpreted the introduction of information technology in the law. Their work shows different interests and concerns.

The focus of Vittorio Frosini's thought is on the ways in which information technology can change the human psychology, social relations, and in particular on the way in which they may experience and practice the law. He views favourably the introduction of technology in the legal domain, but believes that human experience and human cognition cannot fully be captured by it. The humans and their internal consciousness, which is irreducible to computations, must continue to provide the necessary creative input to the artificial world that is created by the humans. This world is in a way the objectivation of human nature, and therefore it is also a part of humanity. Perhaps this view, namely, the unity of human thought and of human creation, reflects the idealistic background of the author.

Mario Losano has a different objective, namely, the way in which computing can provide models of the legal system, as consisting of rules, agents and institutions. In modern terms, we may say that the idea is to combine legal logic, argumentation and agent based-simulation. Such models could be useful not only to support legal practice, but also to improve legal science. Thus we might say that, from Losano's perspective, ICTs play an epistemological role, as new and better ways for theorising on the legal system, understanding its functioning and anticipating its operation.

Lombardi Vallauri is mostly interested in legal reasoning and problem solving, as well as on the role that computing can play in it. Moving from a legal theory that emphasises the moral and political role of legal decision-making (in particular within legal interpretation), his central problem concerns the possibility of capturing those aspects, as well of legal reasoning, through computations. Thus, his problem concerns the possibility to capture also political and ethical choices, which are connected to human consciousness, through technological tools.

Finally, Renato Borruso differs from all previous authors, since he focuses on the way in which ICT can be used for a different way of regulating society, which departs from human limitation and relies on the extensive use of algorithms. He does not view ICT's inability to capturing human knowledge and reasoning as a limitation, but rather as an advantage: by relying on algorithmic models of the law, we can obtain certainty and efficiency, and accomplish the mission of legal Enlightenment.

It seems to us that the work of these four authors is still relevant today, when the ICT revolution is ongoing, and its impacts on the law are more and more important. Creating a human artificial legal world (Frosini), providing scientifically sound computable models of the legal system (Losano), capturing discretion and choice (Lombardi) and experiencing new non-human approaches to legal issues (Borruso) are key challenges for the future.

References

Allen, L. (1957). Symbolic logic. A razor-edged tool for drafting and interpreting legal documents. *The Yale Law Journal, 66*(6), 833–879.

Ballweg, O. (1965). Science, prudence et philosophie du droit. *Archiv für Rechts- und Sozialphilosophie, 4*, 543–560.

Bobbio, N. (1955). *Studi sulla teoria generale del diritto*. Torino: Giappichelli.

Borruso, R., Di Giorgi, R. M., Mattioli, L., & Ragona, M. (2004). *L'informatica del diritto*. Milano: Giuffrè.

Butler, S. (1863). *Darwin among the machines*. Christchurch: The Press.

Cotta, S. (1966). Il giurista e la società in trasformazione. *Jus, 1*, 7–27.

Frosini, V. (1968). *Cibernetica, diritto e società*. Milano: Edizioni di Comunità.

Knapp, V. (1964). *Théorie du droit et cybernétique, in Etudes juridiques offertes à Léon Julliot de la Morandière* (pp. 233–242). Paris: Dalloz.

Loevinger, L. (1963). Jurimetrics: The methodology of legal inquiry. *Law and contemporary problems, 28*(1), 5–35.

Lombardi Vallauri, L. (1975). Democraticità dell'informazione giuridica e informatica. *Informatica e Diritto, 1*, 1–25.

Lombardi Vallauri, L. (1982). Informatica e criteri "politici" o valutativi della decisione giuridica. *Jus, 3*, 303–315.

Lombardi Vallauri, L. (1995). Verso un sistema esperto giuridico integrale. Esempi scelti dal diritto dell'ambiente e della salute. In C. Ciampi, F. S. Natali, & G. T. Elmi (Eds.), *Atti del Convegno venticinquennale dell'I.D.G. I* (pp. 3–18). Padova: CEDAM.

Lombardi Vallauri, L. (2002). *Riduzionismo e oltre. Dispense di filosofia per il diritto*. Padova: CEDAM.

Lombardi Vallauri, L. (2014). Prefazione. Un'informatica giusliberista, democratica, filosofica. In G. Peruginelli & M. Ragona (Eds.), *L'informatica giuridica in Italia. Cinquant'anni di studi, ricerche ed esperienze* (pp. 13–24). Napoli: Edizioni Scientifiche Italiane.

Lombardi Vallauri, L., & Trautteur, G. (1969). Giurisprudenza e cibernetica. *Rivista Internazionale di Filosofia del Diritto, 4*, 423–439.

Losano, M. G. (1994). Informatica giuridica. In *Enciclopedia delle scienze sociali* (pp. 711–719). Roma: Istituto dell'Enciclopedia Italiana Treccani.

Marcuse, H. (1964). *One-dimensional man. Studies in the ideology of advanced industrial society*. Boston: Beacon Press.

Merleau-Ponty, M. (1964). *L'Œil et l'Esprit*. Paris: Gallimard.

Micheal, D. N. (1962). *Cybernation: The silent conquest*. Santa Barbara: Center for the Study of Democratic Institutions.

Philipps, L. (1993). Von der hellen zur dunklen Seite des computers. In M.-T. Tinnefeid, L. Philipps, & K. Weis (Eds.), *Die dunkle Seite des chips. Herrschaft und Beherrschbarkeit neuer Technologien* (pp. 3–15). München-Wien: Oldenbourg.

Somenzi, V. (1993). Cibernetica, informatica e filosofia della scienza. In *La cultura informatica in Italia. Riflessioni e testimonianze sulle origini 1950-1970* (pp. 161–190). Torino: Bollati Boringhieri.

Wiener, N. (1948). *Cybernetics, or control and communication in the animal and the machine*. Cambridge: The MIT Press.

Wiener, N. (1950). *The human use of human beings. Cybernetics and society*. Boston: Houghton Mifflin Company.

Chapter 8
On the Principle of Privacy by Design and Its Limits: Technology, Ethics and the Rule of Law

Ugo Pagallo

Abstract Over the last years lawmakers, privacy commissioners and scholars have discussed the idea of embedding data protection safeguards in ICT and other types of technology, by means of value-sensitive design, AI and legal ontologies, PeCAM platforms, and more. Whereas this kind of effort is offering fruitful solutions for operating systems, health care technologies, social networks and smart environments, the paper stresses some critical aspects of the principle by examining technological limits, ethical constraints and legal conditions of privacy by design, so as to prevent some misapprehensions of the current debate. The idea should be to decrease the entropy of the system via "digital air-bags" and to strengthen people's rights by widening the range of their choices, rather than preventing harm generating behaviour from occurring through the use of self-enforcement technologies.

Keywords Artificial intelligence and law · Data protection · Information ethics · Legal ontologies · Rule of law · Privacy by design

---

This chapter was originally published as Pagallo U. (2012) On the Principle of Privacy by Design and its Limits: Technology, Ethics and the Rule of Law. In: Gutwirth S., Leenes R., De Hert P., Poullet Y. (eds) European Data Protection: In Good Health? Springer, Dordrecht.

U. Pagallo (✉)
Università degli Studi di Torino, Torino, Italy
e-mail: ugo.pagallo@unito.it

S. Chiodo, V. Schiaffonati (eds.), *Italian Philosophy of Technology*, Philosophy of Engineering and Technology 35,
https://doi.org/10.1007/978-3-030-54522-2_8

8.1 Introduction

In the first edition of *The Sciences of Artificial* (1969), Herbert A. Simon lamented the lack of research on "the science of design" which characterized the curricula of both professional schools and universities throughout three decades after the Second World War. In the phrasing of the Nobel laureate, the reason hinged on academic respectability, because "in terms of the prevailing norms, academic respectability calls for subject matter that is intellectually tough, analytic, formalizable, and teachable. In the past much, if not most, of what we knew about design and about artificial sciences was intellectually soft, intuitive, informal, and cook-booky" (Simon 1996, p. 112).

Thirty years later, in *Codes and Other Laws of Cyberspace* (1999), Lawrence Lessig similarly stressed the lack of research on the impact of design on both social relationships and the functioning of legal systems, that is, how human behaviour may be shaped by the design of spaces, places and artefacts.

Thenceforth, the scenario has dramatically changed. Not only, according to Simon, an academically respectable "science of design" has emerged since the mid 1970s, when the Design Research Centre was founded at Carnegie Mellon University (the institute became the "Engineering Design Research Centre" in 1985). Significantly, over the last 10 years, legal scholars and social scientists have increasingly focused on the ethical and political implications of employing design mechanisms to determine people's behaviour through the shaping of products, processes, and Information & Communication Technology (ICT)-interfaces and platforms.

On one hand, let me mention work on the regulatory aspects of technology in such fields as universal usability (Shneiderman 2000); informed consent (Friedman et al. 2002); crime control and architecture (Katyal 2002, 2003); social justice (Borning et al. 2004); allegedly perfect self-enforcement technologies on the internet (Zittrain 2007); and design-based instruments for implementing social policies (Yeung 2007).

On the other hand, following seminal work on the ethics of design (Friedman 1986; Mitcham 1995; Whitbeck 1996), and privacy (Agre 1997), it is noteworthy that scholars have examined data protection issues raised by the design of ICT, by the means of value-sensitive design (Friedman and Kahn 2003; Friedman et al. 2006), legal ontologies (Abou-Tair et al. 2006; Mitre et al. 2006; Lioudakis et al. 2007), projects on platforms for privacy preferences (P3P), (Jutla and Zhang 2005; Cranor et al. 2008; Reay et al. 2009) and PeCAN platforms (Jutla et al. 2006; Jutla 2010), down to the topology of complex social networks (Pagallo 2007). In addition, the idea of incorporating data protection safeguards in ICT was the subject matter of both "Privacy by Design. The Definitive Workshop" organized in Madrid in November 2009 (Cavoukian 2010), and the "Intelligent Privacy Management Symposium" held at Stanford University, CA, on 22–24 March 2010 (the program is online at http://research.it.us.edu.au/magic/privacy2010/schedule.html).

Although the idea of embedding privacy safeguards in information systems and other types of technology is not new, for example, recital 46 and Article 17 of the European Union (EU) directive from 1995 (D-46/95/EC), privacy commissioners have been particularly active in recent times. For example, in the document on *The Future of Privacy* from the 1 December 2009, the European authorities on data protection, that is, the EU Working Party Article 29 D-95/46/EC have frankly admitted that a new legal framework is needed and, more particularly, it "has to include a provision translating the currently punctual requirements into *a broader and consistent principle of privacy by design*. This principle should be binding for technology designers and producers as well as for data controllers who have to decide on the acquisition and use of ICT" (Working Party (WP) Article 29 D-95/46/EC 2009). Among the examples of how the new principle can contribute to better data protection, theWP29 recommends what Norman Potter presented in his 1968 book *What is a Designer* (Potter 2002) as "environmental design" (i.e. designing spaces) and "product design" (i.e. forging objects).

As an illustration of the first kind of design, think about people's anonymity and the challenge of protecting people's privacy in public (Nissenbaum 1998). While the use of, say, CCTVs proliferates and seems unstoppable, the European authorities on data protection propose to design video surveillance in public transportation systems, in such a way that faces of individuals cannot be recognizable (Working Party (WP) Article 29 D-95/46/EC 2009).

Similarly, when making personal data anonymous is considered a priority, matters of design also concern how we organize data processes and product design. A typical instance is given by the WP29's example on the processing of patient names in hospitals via information systems, where patient names should be kept separated from data on medical treatments or health status. Likewise, in accordance with the principle of controllability and confidentiality of the data to be processed, biometric identifiers "should be stored in devices under control of the data subjects (i.e. smart cards) rather than in external data bases" (Working Party (WP) Article 29 D-95/46/EC 2009).

(In the third section of the paper, I address another kind of design that Norman Potter calls communication design. A good example is given by the user friendliness of ICT interfaces and public complaints against Facebook's data protection policies. Whether or not we buy this form of privacy by design, the social network announced on 26 May 2010 to have "drastically simplified and improved its privacy controls" which previously amounted to 170 different options under 50 data protection-related settings...)

Meanwhile, the Ontario's Privacy Commissioner, Ann Cavoukian, has developed the formula "privacy by design" she invented in the late 1990s, so as to cope with the "ever-growing and systemic effects" of both ICT and large-scale networked data systems (Cavoukian 2009). After more than 10 years of efforts and increasing success, the Commissioner organized the aforementioned "definitive workshop" on the principle of privacy by design in November 2009. On that occasion, Cavoukian summed up the idea of handling today's data protection issues, according to seven principles:

1. We have to view data protection in proactive rather than reactive terms, making privacy by design preventive and not simply remedial;
2. Personal data should be automatically protected in every IT system as its default position;
3. Data protection should accordingly be embedded into design;
4. The full functionality of the principle which follows from (2) and (3) allows a positive-sum or win-win game, making trade-offs unnecessary (e.g. privacy vs. security);
5. A cradle-to-grave, start-to-finish, or end-to-end lifecycle protection ensures that privacy safeguards are at work even before a single bit of information has been collected;
6. No matter the technology or business practices involved, the design project should make data protection mechanisms visible and transparent to both IT users and providers;
7. Finally, the principle "requires architects and operators to keep the interests of the individual uppermost by offering such measures as strong privacy defaults, appropriate notice, and empowering user-friendly options" (Cavoukian 2010). In other words, privacy by design requires an individual-focused respect for user privacy.

In the light of these seven tenets, I admit that the principle of privacy by design looks particularly promising in such different fields as data protection in CCTV systems, biometrics, social networks, smart environments, data loss prevention and more. The principle may in fact represent a turning point in how we address most of the challenges in data protection due to the development of cloud computing, the internet of things, or the semantic Web (Kim et al. 2002; Jutla and Xu 2004; Breuker et al. 2009), by strengthening people's habeas data and allowing us to prevent the risk of hampering economic growth due to alleged privacy reasons. Moreover, the principle shows an effective way to solve some of the extra-territorial legal effects and jurisdictional issues created by digital technology, since privacy assurance can become a default mode of operation for both private companies and public institutions in transnational law (Pagallo 2008).

However, this success entails its own risks, such as current misapprehensions in today's debate and divergent interpretations of the principle among commissioners and scholars. Whereas some propose a version of the principle of privacy "as" design, that is, making most legal provisions on data protection preventive and automatic, it is far from clear what type of design mechanism the WP29 is referring to, when claiming that privacy by design "should be binding" for data controllers, technology designers and producers (Working Party (WP) Article 29 D-95/46/EC 2009). Should the aim be to integrate compliance with regulatory frameworks through design policies or, conversely, should the aim be to prevent harm-generating behaviour from occurring?

In order to offer a hopefully comprehensive view of these issues, this chapter is presented in three sections.

First, I examine the idea of making all the legal provisions on data protection automatic, according to points (ii), (iii), and (v) of Cavoukian (2010). As shown by 10 years of efforts on the development of platforms for privacy preferences, "the P3P specification is not yet mature enough in terms of element definitions to handle many legal subtleties cleanly" (Jutla 2010). Far from being mere subtleties, however, the first section of the chapter aims to show that such legal hurdles to the "end-to-end lifecycle" of data protection regard some of the most important notions of the legal framework, that is, highly context-dependent normative concepts like data controller, security measure or, even, personal data.

Secondly, these difficulties emphasize the ethical issues of design and the strong moral responsibilities behind the use of alleged perfect self-enforcement technologies. Whereas individual preferences play a crucial role in determining levels of access to and control over information in digital environments, people's behaviour would unilaterally be determined on the basis of automatic techniques rather than by choices of the relevant political institutions (Lessig 2004). In the name of individual autonomy, this is why I propose to frame the ethical issues of design and its modalities, by adopting a stricter version of the principle (Pagallo 2009).

Thirdly, such a stricter version of privacy by design is examined in connection with the democratic rule of law and the principle that individuals have to have a say in the decisions affecting them. As suggested by the European Data Protection Supervisor (EDPS), Peter Hustinx, in the Opinion from 25 July 2007 (2007/C 255/01), the challenge of protecting personal data "will be to find practical solutions" through typical transnational measures such as "the use of binding corporate rules by multinational companies" and "international agreements on jurisdiction" (*op. cit.*, § 44). Analogously, efforts should aim at "promoting private enforcement of data protection principles through self-regulation and competition" (*op. cit.*, § 65), while "accepted standards such as the OECD-guidelines for data protection (1980) and 153 UN-Guidelines could be used as basis" (*op. cit.*, § 44).

To conclude, privacy by design should encourage people to change their conduct (e.g. user-friendly interfaces), or limit the effects of harmful behaviour (e.g. security measures) by strengthening people's rights and broadening the range of their choices. There is, indeed, "respect for user privacy" (Cavoukian 2010), when averting both the risks of paternalistic drifts and further conflicts of values in the realm of privacy by design. Rather than a "cradle-to-grave lifecycle" of automatic protection, let us reinforce the pre-existing individual autonomy (Pagallo 2011a).

8.2 Technology and Its Limits

I mentioned some of the different ways the scholars have addressed points (ii), (iii) and (v) of Cavoukian's scheme, so that personal data should automatically be protected in every IT system as its default position and even before a bit of information has been collected. Leaving aside value sensitive design-approaches, P3P and PeCAN platforms, let me focus on current efforts in Artificial Intelligence (AI) & Law and, more specifically, in legal ontologies, so as to stress the first limit of the principle of privacy by design, that is, current state-of-the-art in technology.

Legal ontologies is the field of AI that aims to model concepts traditionally employed by lawyers through the formalization of norms, rights and duties, in fields like criminal law, administrative law, civil law, etc. (Breuker et al. 2009; Casanovas et al. 2010). The objective is that even a machine should comprehend and process this very information, by preliminarily distinguishing between the part of the ontology containing all the relevant concepts of the problem domain through the use of taxonomies (e.g. ontological *requirements*), and the ontology which includes both the set of rules and restraints that belong to that problem domain (e.g. ontological *constraints*). An expert system should thus process the information in compliance with regulatory legal frameworks through the conceptualization of classes, relations, properties and instances pertaining to that given problem domain of data protection. Following what has been said about the ongoing project on the "Neurona Ontology" developed by Pompeu Casanovas and his research team in Barcelona, Spain, the goal is to implement new technological advances in managing personal data and provide organizations and citizens "with better guarantees of proper access, storage, management and sharing of files" (Casellas et al. 2014). By programming the software of the system to comply with regulatory frameworks of data protection, it is feasible to help company officers and citizens "who may have little or no legal knowledge whatsoever."

In technical terms, we should pay attention to the bottom-up approach that starts from legal concepts defined by scholars. A traditional top-down approach works well for the *topmost level*, where the representation instruments are at the disposal of the ontology-builders and the basic conceptual primitives such as relation, role, qualia, processes, etc., are precisely defined. However, a lot of issues arise when the *core ontology* level is taken into account, because the amount of information involved in the project of making data protection safeguards automatic is hardly compressible. Simply put, data protection regulations not only include "top normative concepts" such as notions of validity, obligation, prohibition, and the like. These rules present also highly context-dependent normative concepts like personal data, security measures, or data controllers. In order to grasp some of the difficulties of embedding data protection safeguards in a software program, simply reflect on three facts:

1. In the aforementioned document on *The Future of Privacy*, the EUWP29 warns that "Web 2.0 services and cloud computing are blurring the distinction between data controllers, processors and data subjects";
2. In the Opinion from the 1 February 2010, the EUWP29 insisted that "the concept of controller is a functional concept, intended to allocate responsibilities where the factual influence is, and thus based on a factual rather than a formal analysis. Therefore, determining control may sometimes require an in-depth and lengthy investigation" (doc. 00264/10/EN WP 169)

3. Finally, on 23 March 2010, the European Court of Justice declared that liability of online referencing service providers depends on "the actual terms on which the service is supplied." In other words, according to the judges in Luxembourg, it is necessary to determine "whether the role played by that service provider is neutral, in the sense that its conduct is merely technical, automatic and passive, pointing to a lack of knowledge or control of the data which it stores" (*Google v. Louis Vuitton case*, § 114 of the decision).

The difficulty to program the WP29's "factual influence" of the data controller or the ECJ's "actual terms" of the service provided on the internet does not mean that projects on legal ontologies and privacy by design should be abandoned. On the contrary, these difficulties suggest a bottom-up rather than a top-down approach, in order to lawfully process growing amounts of personal data. By splitting the work into several tasks and assigning each to a working team, we should start from smaller parts and sub-solutions of the design project, to end up with "global answers" (Casellas et al. 2014). The evaluation phase consists in testing the internal consistency of the project and, according to Herbert Simon's "generator test-cycle," entails the decomposition of the complete design into functional components. The test generates alternatives and examines them against the set of *requirements* and *constraints*, so that "important indirect consequences will be noticed and weighed. Alternative decompositions correspond to different ways of dividing the responsibilities for the final design between generators and tests" (Simon 1996, p. 128).

Further criteria and empirical methods have been proposed: apart from functional efficiency, we should consider the robustness, reliability, and usability of design projects. Evaluation and verification of the design can additionally employ automated and regression-oriented tests, use of prototypes, internal checks among the design team, users tests in controlled environments, surveys, interviews and more (Flanagan et al. 2008). On this basis, we can quantify the growing amount of personal data processed in compliance with regulatory frameworks. This is the focus of the research on legal ontologies and the support of privacy preservation in location-based services (Mitre et al. 2006), the management of information systems (Abou-Tair et al. 2006; Casellas et al. 2014), or middleware architectures for data protection (Lioudakis et al. 2007), each of which aims at integrating smaller parts and sub-solutions into the design. Remarkably, there are even cases where the conceptualization of classes, relations, properties and instances pertaining to a given problem domain does not seem particularly complex, for example, the design of information systems for hospitals to ensure that patient names are kept separated from data on medical treatments or health status (Working Party (WP) Article 29 D-95/46/EC 2009).

However, by lawfully processing growing amounts of personal data, it does not follow that goals (ii), (iii) and (v) of Cavoukian's scheme, that is, making data protection automatic by design, are at hand. Besides the difficulty of formalizing highly context-dependent concepts such as data processor or data controller, designers must take into account that privacy is not a zero-sum game between multiple instances of access to and control over the main role when individuals modulate different levels of access and control, depending on the context and its circumstances (Nissenbaum 2004). Moreover, people may enjoy privacy in the midst of a crowd and without having total control over their personal data, whereas total control over that data does not necessarily entail any guarantee of privacy (Tavani 2007). Such constraints emphasize the first limit of the principle: in accordance with today's state-of-the-art, no expert system allows us to fully achieve goals (ii), (iii) and (v) of Cavoukian's principles of privacy by design. To the best of my knowledge, it is impossible to programme software so as to prevent, say, forms of harm-generating behaviour as simple as defamations, but leaving aside technical details, how about the desirability of such a project?

8.3 Ethical Constraints

Some of the most relevant problems concerning today's data protection hinge on the information revolution and the fact that no clear legal boundaries exist in digital environments. State-action is often ineffective due to the ubiquitous nature of information: while citizens of nation states are often affected by conduct that the state is unable to regulate (e.g. spamming), this situation may also lead to the illegitimate condition where a state claims to regulate extraterritorial conduct by imposing norms on individuals, who have no say in the decisions affecting them (Post 2002). According to the 2007 EDPS Opinion, "this system, a logical and necessary consequence of the territorial limitations of the European Union, will not provide full protection to the European data subject in a networked society where physical borders lose importance [...]: the information on the Internet has an ubiquitous nature, but the jurisdiction of the European legislator is not ubiquitous" (Hustinx 2007).

The ineffectiveness of state-action depends on how ICT allows information to transcend traditional legal borders, questioning the notion of the law as made of commands enforced through physical sanctions. Spamming is again a good example for it is par excellence *transnational* and does not diminish despite harsh criminal laws (as the *CAN-SPAM Act* approved by the U.S. Congress in 2003). Since the mid 1990s, as a consequence, companies and big business have tried to find out a remedy for the apparent inefficacy of state-action in protecting their own rights. While lobbying national and international law-makers in the copyright field, some of the most relevant companies focused on how to enforce their (alleged) exclusivity rights through the development of self-enforcement technologies, for example, Digital Rights Management (DRM). By enabling right-holders to monitor and regulate the use of their own copyright protected works, companies would have prevented unsolvable problems involving the enforceability of national laws and conflicts of law at the international level.

However, whether or not DRM works and can be held to be fair, the aim of privacy by design, that is, to exert "automatic control" over personal information is even more debatable than the use of DRM technology for the protection and enforcement of digital copyright. Whereas Jobs (2007) conceded in his *Thoughts on Music*

(2007) that DRM-compliant systems raise severe problems of interoperability and, hence, antitrust-related challenges, the aim of privacy by design to automatically prevent harm-generating conducts from occurring looks problematic for three reasons.

First, we have evidence that "some technical artefacts bear directly and systematically on the realization, or suppression, of particular configurations of social, ethical, and political values" (Flanagan et al. 2008). Specific design choices may result in conflicts between values and, vice versa, conflicts between values may impact on the features of design. Consider the different features that privacy by design acquires, once data protection is grasped in terms of property rights or human dignity, of total control or contextual integrity, of restricted access to or limited control over digital information. At the end of the day, should an artefact be designed in accordance with the opt-in model for users of electronic communication systems or, vice versa, according to the opt-out approach? Moreover, reflect upon the information system of hospitals which I mentioned in the previous section: should we privilege the efficacy and reliability of that information system in keeping patient names separated from data on medical treatments or health status? How about users, including doctors, who may find such mechanism too onerous?

Secondly, attention should be drawn to the difficulties of achieving such total control. Doubts are cast by "a rich body of scholarship concerning the theory and practice of 'traditional' rule-based regulation [that] bears witness to the impossibility of designing regulatory standards in the form of legal rules that will hit their target with perfect accuracy" (Yeung 2007). The worthy aim to prevent people's privacy infringements involves *strong moral responsibility* of both the designers and public authorities, in that use of self-enforcement technologies collapses "the public understanding of law with its application eliminating a useful interface between the law's terms and its application" (Zittrain 2007). As a response to the inefficacy of state-action in digital environments, the development of this type of technology risks to curtail freedom and individual autonomy severely, because people's behaviour would unilaterally be determined on the basis of technology, rather than by choices of the relevant political institutions. In the phrasing of Larry Lessig, "the controls over access to content will not be controls that are ratified by courts; the controls over access to content will be controls that are coded by programmers" (Lessig 2004).

Finally, there is the issue of security in balancing different levels of access and control via software: the expert system should not be capable to only balance personal preferences *and* matters of "property rights" (Spinello 2003), "trade-offs" (Nissenbaum 2004), or "integrity" (Grodzinsky and Tavani 2008), which often depend on contextual choices. In fact, design projects should be capable to evaluate this (sensitive) information safely, although experts warn that "the only truly secure system is one that is powered off, cast in a block of concrete and sealed in a lead-lined room with armed guards—and even then I have my doubts" (Garfinkel and Spafford 1997). Whereas the use of self-enforcement technologies may be compatible with the precautionary principle in the area of intellectual property rights (Weckert and Moor 2004; Clarke 2005), this does not seem to be the case of privacy by design. DRM's shortcomings principally impact on companies that employ such devices to protect their own copyright protected files: in the case of privacy, however, the use of alleged self-enforcement technologies would directly impact on everyone of us as "informational objects" (Floridi 2006).

Therefore, I suggest abandoning the idea of making data protection automatic by design, so as to prevent every harm-generating conduct from occurring.

Rather, we should focus on other mechanisms we may aim at through design, that is, both the aim to encourage the change of people's behaviour via user friendly interfaces and to decrease the impact of harm-generating conducts through "digital air-bags" as encryption and other security measures (von Ahn et al. 2008). Let us examine projects on privacy by design, by considering today's "habeas data" in connection with the principle of the rule of law.

8.4 Habeas Data

As shown by the proceedings of the 2009 IVR 24th World Congress in Beijing, China, on *Global Harmony and the Rule of Law* (see http://www.ivr2009.com/), not only "harmony" is a very controversial concept of the millennial political tradition of China, but Western scholars are sparkly divided by the meaning of the "rule of law" as well (whether or not we conceive it as the German *Rechtsstaat*, the French *État de droit*, the Spanish *Estado de derecho*, the Italian *Stato di diritto*, and so forth). While the appeal of the formula historically derives from Plato's distinction between the "empire of the laws," that is, rule by law, and "empire of men," namely, rule under the will of men, it is unclear whether the view of the rule of law adopts a thin-procedural or a thick-substantive approach to distinguishing between rule of law, rule by law, etc. (Summers 1993). It is noteworthy that "despite a growing empirical literature, there remain serious doubts about the relationship, and often causal direction, between the rule of law and the ever-increasing list of goodies with which it is associated, including economic growth, poverty reduction, democratization, legal empowerment and human rights" (Peerenboom 2009).

In this context, it suffices to dwell on the traditional connection between the rule of law and the principle of *habeas corpus*, that is, individual protection against arbitrary (both public and private) action. Over the last two decades, several provisions on data protection, for example, Article 8 of the EU Charter of Fundamental Rights, have complemented the traditional version of the principle of *habeas corpus*, linked to the physical body, with a new kind of protection, that is *habeas data*, as an extension of that protection to the electronic body of each individual (Rodotà 2006). What is at stake with the principle of privacy by design accordingly concerns whether some kinds of "electronic protection" violate people's right to have a say in the decisions affecting them, that is, what the German Constitutional Court frames in terms of the individual "informational self-determination." As well known, the *Bundesverfassungsgericht* (BVerfG) has furthered the concept since its *Volkszählungs-Urteil* ("census decision") from 15 December 1983.

Furthermore, we have to ascertain whether protection of the electronic body via design policies may entail what Kant criticized as paternalism (Kant 1891). By adopting a sort of automatic *habeas data*, the threat is to impinge on the "property of the will" to rule over itself, so that, according to *Grounding for the Metaphysics of Morals*, the risk is to overprotect individuals against every harm and, even, against themselves. In the light of the panoply of projects and approaches in the field of data protection mentioned in this paper, it is crucial to preliminarily distinguish three aims of design (Yeung 2007), so as to take sides on the legal constraints of the principle:

1. Design may prevent harm-generating behaviour from occurring;
2. Design may aim to decrease the impact of harm-generating conducts;
3. Design may encourage the change of social behaviour.

Although design is not necessarily digital (Lessig 1999), the information revolution has induced a more sophisticated legal enforcement than, say, the installation of speed bumps in roads to reduce the velocity of cars. In the case of data protection, scholars should determine what kind of design mechanism is compatible with the tenets of the rule of law, in order to ensure the minimization and quality of the data, its controllability, transparency, and confidentiality, down to the user friendliness of information interfaces.

The first aim of design mechanism, that is, the prevention of harmful conducts thanks to the use of self-enforcement technologies, seems highly problematic in this context. Besides the technical and ethical reasons that make such a protection neither feasible nor desirable in the realm of *habeas data*, perfect automation of data protection mechanisms impinges on the individual right to the "informational self-determination"—that is, the *informationelle Selbstbestimmung* of the BVerfG—which includes the right to determine whether personal data can be collected and, eventually, transmitted to others; the right to determine how that data may be used and processed; the right to access that data and, where necessary, to keep it up to date; besides the right to delete that data and to refuse at any time to have the data processed. Since the enforcement and guarantee of most of these rights are beyond today's state-of-the-art in technology, it follows that an automatic *habeas data* would impose norms on subjects who have no say in the decisions affecting them (Lessig 2004; Zittrain 2007), thereby making people lose their capacity for making moral choices (Brownsword 2005). Instead of letting people determine autonomously levels of access to and control over personal data, depending on personal choices and circumstances, the use of self-enforcement technologies seems incompatible with a basic tenet of the democratic rule of law—autonomy.

But, how about the remaining mechanisms of privacy by design, that is, when the aim is not to prevent certain actions from being chosen overall, but to merely inspire a different conduct by encouraging people to change their behaviour or decreasing the impact of harm-generating conducts? Are these aims compatible with the rule of law?

On one hand, design mechanisms closely regard point (vii) of Cavoukian's principles of privacy by design, that is, the individual-focused approach respectful of user privacy. The idea is well represented by current efforts on security measures, location-based services, friendly interfaces, P2P overlay platforms, default settings and more (Pagallo 2011b). In all the examples of *this* type of design mechanisms, it is arguably correct to stress that "privacy assurance must ideally become an organization's default mode of operation" (Cavoukian 2009). The aim to decrease the impact of harm-generating conducts, as air-bags do in cars, does not seem to impinge on individual autonomy and personal data, because ICT mechanisms as well as air-bags are designed to respect people's choices when they, say, drive cars or modulate different levels of privacy, according to the context. As an instance of "digital air-bags," consider "the power of defaults" (Kesan and Shah 2006), so that we can ensure that values of design are appropriate for novice users and, still, the system improves efficiency. Likewise, reflect on modifications to user interfaces by increasing, or reducing, the prominence of a default setting, so as to allow users to configure and use their software as they deem appropriate. Moreover, consider security measures, such as reCAPTCHA, that aim to prevent automated programs from abusing online services (von Ahn et al. 2008). The aim of such design projects that reduce the effects of harmful conducts fully respects the Kantian principle of autonomy because the only responsibility, both legal and moral, which is at stake with this type of design mechanism concerns the technical meticulousness of the project and its reliability, as it occurs with security measures for the information systems of an atomic plant or a hospital.

On the other hand, by encouraging the change of social behaviour, design projects suggest to assess the impact of design choices on people's conduct. This is the case of the free-riding phenomenon on P2P file-sharing networks, where most peers tend to use these systems to find information and download their favourite files without contributing to the performance of the system. Whilst this behaviour is triggered by many properties of P2P applications like anonymity and hard traceability of the nodes, designers have proposed ways to tackle the issue through incentives based on trust (e.g. reputation mechanisms), trade (e.g. services in return), or alternatively slowing down the connectivity of the user who does not help the process of file-sharing (Glorioso et al. 2010). The editorials in *The Economist* aside, some scholars have nevertheless stressed a threat of paternalism behind the very idea of encouraging the change of people's behaviour (Kuner 2003; Volkman 2003). After all, this type of design mechanism may represent a way of modelling social conduct so as to protect people against all forms of harm. This threat makes urgent a normative view point such as information ethics (Floridi 2005), online privacy policies (Tavani 2007), ethics of design (Friedman 1986; Mitcham 1995; Whitbeck 1996; Flanagan et al. 2008), machine ethics (Moor 2006; McLaren 2006), and more, for we should previously test the goodness of data protection laws, in order to prove the goodness of our own design projects. Is there a way to ensure that privacy by design does not violate the anti-paternalistic stance of the rule of law by encouraging people to change their conduct? How about conflicts between values that necessarily reverberate on design choices? Is, say, Jeffrey Rosen right, when stressing the fear that "cultural differences will make thoughtful regulation difficult" in data protection? What does it mean for data protection that "the French will bare their breasts but not their salaries and mortgages, and the reverse is true in the U.S."? (As Rosen declares in Mills 2008).

Although it is notoriously difficult to solve conflicts of values with their divergent interpretations, we might prevent most issues in today's *cul de sac* by embracing one of the several examples and design mechanisms put forward by the EU Working Party's document on *The Future of Privacy*. Whether or not you agree that the European legal framework "is clearly and deeply flawed as an account of what informational protection is all about" (Volkman 2003), we need not sympathize with Brussels to follow the proposal that the principle of privacy by design should be implemented in accordance with a bottom-up rather than a top-down approach, that is, depending on individual autonomous choices via self-regulation and competition among private organizations (Working Party (WP) Article 29 D-95/46/EC 2009).

As a result, besides a stricter version of privacy by design as a way to decrease the "informational entropy" of the system through "digital air-bags," we find a further design mechanism compatible with the rule of law. When encouraging people to change their behaviour by the means of design, the overall goal should be to reinforce people's pre-existing autonomy, rather than building it from scratch. In the wording of the EU privacy commissioners, the principle should enable business and individuals to "take relevant security measures by themselves" (Working Party (WP) Article 29 D-95/46/EC 2009).

8.5 Conclusions

It is unlikely that privacy by design will offer the one-size-fits-all solution to the problems in the realm of data protection, although privacy by design is a good candidate for understanding how we have coped with privacy issues over the last few years. The principle may in fact be a turning point in how we address most of the challenges in data protection, by strengthening people's habeas data, without hampering economic growth for alleged privacy reasons. In different fields as data protection in CCTV systems, biometrics, social networks, smart environments, data loss prevention and more, projects are increasingly processing growing amounts of personal data in compliance with current normative frameworks, strengthened by the capacities of computers to draw upon the tools of AI and operations research.

Notwithstanding the merits, however, there are three reasons why we should be aware of the limits of privacy by design. These limits are especially relevant when the aim is to automatically protect personal data as the default position of every ICT system, even before a single bit of information has been collected, that is, points (ii), (iii) and (v) of Cavoukian's scheme on the principle (Cavoukian 2010). Let me sum up these limits.

First, I mentioned work on legal ontologies, value-sensitive design, P3P or PeCAN platforms, so as to show the limits of today's state-of-the-art in technology. Besides the difficulty of modelling highly context-dependent normative concepts as data controllers and the "neutrality" of the services provided on the internet, designers should take into account that privacy is not a zero-sum game but concerns personal choices on levels of access to and control over information that often depend on the context. Making all the provisions of data protection automatic is simply out of reach.

The second limit involves the ethical constraints of the approach and the process of both law-making and legal enforcement. Not only conflicts between values do impact on the features of design but, vice versa, design choices may result in further conflicts between values. Since privacy may be conceived in terms of human dignity or property rights, of contextual integrity or total control, it follows that privacy by design acquires many different features. Moreover, self-enforcement technologies risk to curtail freedom and individual autonomy severely, because people's behaviour would be determined on the basis of design rather than by individual choices.

Finally, two tenets of the rule of law, that is, autonomy and anti-paternalism, stressed the legal constraints of privacy by design as a means to prevent harm generating behaviour from occurring. By adopting a sort of automatic *habeas data*, the risk is to impinge on what Kant defined the "property of the will" to rule over itself and, two centuries later, the BVerfG presented as the individual right to "informational self-determination." Leaving aside the technical unfeasibility of goals (ii), (iii) and (v) of Cavoukian's model, it is undisputable that the more personal choices are wiped out by automation, the bigger the threat of modelling social conduct via design, that is, Kantian paternalism.

As a consequence, this chapter has proposed a stricter version of the principle of privacy by design which seems to be technically feasible, ethically sound and lawful. On one hand, in accordance with goals (i) and (vi) of Cavoukian (2010), privacy by design can legitimately aim to automatically reduce the impact of harm-generating behaviour, so that "privacy assurance must ideally become an organization's default mode of operation" (Cavoukian 2009). Besides values of design that are appropriate for novice users and, hence, procedural constraints for changing the setting of the interfaces on voluntary and fully informed basis, I mentioned security measures that aim to prevent automated programs from abusing online services: "digital air-bags" as friendly interfaces, P2P overlay platforms or default settings will not impinge on individual autonomy, no more than traditional air-bags affect how people drive. On the other hand, in connection with point (vii) of Cavoukian's model, privacy by design can legitimately aim to encourage the change of social behaviour if, and only if, the goal is to strengthen people's rights by widening the range of their choices. This is the version of the principle put forward by the example of both the WP29 and the European Data Protection Supervisor, when endorsing the enforcement of data protection through self regulation and competition (Hustinx 2007; Working Party (WP) Article 29 D-95/46/EC 2009), thus preventing claims of paternalism by fostering individual habeas data.

The result is a final convergence over the "full functionality" of the principle, that is, point (iv) of Cavoukian's scheme. A positive-sum or win-win game becomes possible by embedding data protection safeguards in technology with the aim to encourage people to change their conduct as well as to decrease the effects of harmful behaviour. As shown by current work on legal ontologies, middleware architectures for data protection, the management of information systems, and more, trade-offs such as privacy vs. business, privacy vs. security, privacy vs. copyright, are not always necessary. However, it is not only a matter of business and security—privacy by design concerns a basic tenet of the rule of law such as the principle of autonomy.

References

Abou-Tair, D., el Diehn, I., & Berlik, S. (2006). An ontology-based approach for managing and maintaining privacy in information systems. In *Lectures notes in computer science* (Vol. 4275, pp. 983–994). Berlin: Springer.

Agre, P. E. (1997). Introduction. In P. E. Agre & M. Rotenberg (Eds.), *Technology and privacy: The new landscape* (pp. 1–28). Cambridge: The MIT Press.

Borning, A., Friedman, B., & Kahn, P. H. (2004). Designing for human values in an urban simulation system: Value sensitive design and participatory design. In *Proceedings of eighth biennial participatory design conference* (pp. 64–67). Toronto: ACM Press.

Breuker, J., Casanovas, P., Klein, M., & Francesconi, E. (2009). *Law, ontologies and the semantic web*. Amsterdam: IOS Press.

Brownsword, R. (2005). Code, control, and choice: Why east is east and west is west. *Legal Studies, 25*(1), 1–21.

Casanovas, P., Ugo, P., Sartor, G., & Ajani, G. (2010). *AI approaches to the complexity of legal systems. Complex systems, the semantic web, ontologies, argumentation, and dialogue*. Berlin: Springer.

Casellas, N., Torralba, S., Nieto, J.-E., Merono, A., Roig, A., Reyes, M., & Casanovas, P. (2014). The neurona ontology: A data protection compliance ontology. In *Paper presented at the intelligent privacy management symposium*. Stanford: Stanford University.

Cavoukian, A. (2009). *Privacy by design*. Ottawa: IPC.

Cavoukian, A. (2010). Privacy by design: The definitive workshop. *Identity in the Information Society, 3*(2), 247–251.

Clarke, S. (2005). Future technologies, dystopic futures and the precautionary principle. *Ethics and Information Technology, 7*(4), 121–126.

Cranor, L. F., Egelman, S., Sheng, S., McDonald, A. M., & Chowdhury, A. (2008). P3P deployment on websites. *Electronic Commerce Research and Applications, 7*(3), 274–293.

Flanagan, M., Howe, D. C., & Nissenbaum, H. (2008). Embodying values in technology: Theory and practice. In J. van den Hoven & J. Weckert (Eds.), *Information technology and moral philosophy* (pp. 322–353). New York: Cambridge University Press.

Floridi, L. (2005). Information ethics, its nature and scope. *Computers and Society, 36*(3), 21–36.

Floridi, L. (2006). Four challenges for a theory of informational privacy. *Ethics and Information Technology, 8*(3), 109–119.

Friedman, B. (1986). Value-sensitive design. *Interactions, 3*(6), 17–23.

Friedman, B., Howe, D. C., & Felten, E. (2002). Informed consent in the Mozilla browser: Implementing value-sensitive design. In *Proceedings of 35th annual Hawaii international conference on system sciences*. Washington, DC: IEEE Computer Society.

Friedman, B., Kahn, P. H., Jr., & Borning, A. (2006). Value sensitive design and information systems. In P. Zhang & D. Galletta (Eds.), *Human-computer interaction in management information systems: Foundations* (pp. 348–372). New York: Armonk.

Friedman, B., & Kahn, P. H. (2003). Human values, ethics, and design. In J. A. Jacko & A. Sear (Eds.), *The human computer interaction handbook* (pp. 1177–1201). Mahwah: Lawrence Erlbaum Associates.
Garfinkel, S., & Spafford, E. (1997). *Web security and commerce*. Sebastopol: O'Reilly.
Glorioso, A., Pagallo, U., & Ruffo, G. (2010). The social impact of P2P systems. In X. Shen, H. Yu, J. Buford, & M. Akon (Eds.), *Handbook of peer-to-peer networking* (pp. 47–70). Heidelberg: Springer.
Grodzinsky, F. S., & Tavani, H. T. (2008). Online file sharing: Resolving the tensions between privacy and property interest. In T. W. Bynum, M. Calzarossa, I. De Lotto, & S. Rogerson (Eds.), *Proceedings of ETHICOMP2008 "living, working and learning beyond technology"* (pp. 373–383). Tipografia Commerciale: Mantova.
Hustinx, P. (2007). Opinion of the European Data Protection Supervisor on the Communication from the Commission to the European Parliament and the Council on the Follow-up of the Work Program for Better Implementation of the Data Protection Directive. *Official Journal of the European Union, 25*, 51.
Jobs, S. (2007). *Thoughts on music*. Retrieved 20 April, 2009, from http://www.apple.com/hotnews/thoughtsonmusic/
Jutla, D. N. (2010). Layering privacy on operating systems, social networks, and other platforms by design. *Identity in the Information Society, 3*(2), 319–341.
Jutla, D. N., Bodorik, P., & Zhan, Y. (2006). PeCAN: An architecture for user privacy and profiles in electronic commerce contexts on the semantic web. *Information Systems, 31*(4-5), 295–320.
Jutla, D. N., & Xu, L. (2004). *Privacy agents and ontology for the semantic web. Americas Conference on information systems*. New York: CUSP.
Jutla, D. N., & Zhang, Y. (2005). Maturing E-privacy with P3P and context agents. In *Proceedings of IEEE International Conference on E-Technology, E-Commerce and E-Service, Hong Kong* (pp. 536–541).
Kant, I. (1891). *Kant's principles of politics, including his essay on perpetual peace. A contribution to political science*. Edinburgh: Clark.
Katyal, N. (2002). Architecture as crime control. *The Yale Law Journal, 111*(5), 1039–1139.
Katyal, N. (2003). Digital architecture as crime control. *The Yale Law Journal, 112*(6), 101–129.
Kesan, J. P., & Shah, R. C. (2006). Setting software defaults: perspectives from law, computer science and behavioural economics. *Notre Dame Law Review, 82*, 583–634.
Kim, A., Hoffman, L. J., & Martin, C. D. (2002). Building privacy into the semantic web: Ontology needed now. Semantic web workshop 2002. Honolulu, Hawaii.
Kuner, C. (2003). *European data privacy law and online business*. Oxford: Oxford University Press.
Lessig, L. (1999). *Code and other laws of cyberspace*. New York: Basic Books.
Lessig, L. (2004). *Free culture: The nature and future of creativity*. New York: Penguin Press.
Lioudakis, G., Koutsoloukasa, E., Tselikasa, N., Kapellakia, S., Prezerakosa, G., Kaklamani, D., & Venieris, I. (2007). A middleware architecture for privacy protection. *The International Journal of Computer and Telecommunications Networking, 51*(16), 4679–4696.
McLaren, B. (2006). Computational models of ethical reasoning: Challenges, initial steps, and future directions. *IEEE Intelligent Systems, 2006*, 29–37.
Mills, E. (2008). To be anonymous or not to be, that is the privacy question: Interview to Jeffrey Rosen. *News blog*. Retrieved 15 October, 2010, from http://news.cnet.com/8301-10784_3-9889255-7.html
Mitcham, C. (1995). Ethics into design. In R. Buchanan & V. Margolin (Eds.), *Discovering design* (pp. 173–179). Chicago: University of Chicago Press.
Mitre, H., Gonzàlez-Tablas, A. I., Ramos, B., & Ribagorda, A. (2006). *A legal ontology to support privacy preservation in location-based services. Lectures notes in computer science* (Vol. 4278, pp. 1755–1764). Berlin: Springer.
Moor, J. (2006). The nature, importance, and difficulty of machine ethics. *IEEE Intelligent Systems, 2006*, 18–21.

Nissenbaum, H. (1998). Protecting privacy in an information age: The problem of privacy in public. *Law and Philosophy, 17*(5-6), 559–596.
Nissenbaum, H. (2004). Privacy as contextual integrity. *Washington Law Review, 79*(1), 119–158.
Pagallo, U. (2007). Small world-paradigm and empirical research in legal ontologies: A topological approach. In G. Ajani, G. Peruginelli, G. Sartor, & D. Tiscornia (Eds.), *The multilanguage complexity of European Law: Methodologies in comparison* (pp. 195–210). Florence: European Press Academic.
Pagallo, U. (2008). *La tutela della privacy negli Stati Uniti d'America e in Europa: Modelli giuridici a confronto*. Milano: Giuffrè.
Pagallo, U. (2009). Privacy e design. *Informatica e Diritto, 1*, 123–134.
Pagallo, U. (2011a). Designing data protection safeguards ethically. *Information, 2*(2), 247–265.
Pagallo, U. (2011b). The trouble with digital copies: A short km phenomenology. In G. J. Morais da Costa (Ed.), *Ethical issues and social dilemmas in knowledge management organizational innovation* (pp. 195–210). Hershey: IGI Global.
Peerenboom, R. (2009). The future of rule of law: The challenges and prospects for the field. *Hague Journal on the Rule of Law, 1*(1), 5–14.
Post, D. G. (2002). Against "against cyberspace". *Berkeley Technology Law Journal, 17*(4), 365–1383.
Potter, N. (2002). *What is a designer*. London: Hyphen Press.
Reay, I., Dick, S., & Miller, J. (2009). A large-scale empirical study on p3p privacy policies: Stated actions vs. legal obligations. *ACM Transactions on the Web, 3*, 2.
Rodotà, S. (2006). The retention of electronic communication traffic data. *Revista d'Internet, dret i política, 3*, 53–60.
Shneiderman, B. (2000). Universal usability. *Communications of the ACM, 43*(3), 84–91.
Simon, H. A. (1996). *The sciences of the artificial*. Cambridge: The MIT Press.
Spinello, R. A. (2003). The future of intellectual property. *Ethics and Information Technology, 5*(1), 1–16.
Summers, R. S. (1993). A formal theory of rule of law. *Ratio Iuris, 6*(2), 127–142.
Tavani, H. T. (2007). Philosophical theories of privacy: implications for an adequate online privacy policy. *Metaphilosophy, 38*(1), 1–22.
Volkman, R. (2003). Privacy as life, liberty, property. *Ethics and Information Technology, 5*(4), 199–210.
von Ahn, L., Maurer, B., McMillen, C., Abraham, D., & Blum, M. (2008). reCAPTCHA: Human-based character recognition via web security measures. *Science, 321*(5895), 1465–1468.
Weckert, J., & Moor, J. (2004). Using the precautionary principle in nanotechnology policy making. *Asia Pacific Nanotechnology Forum News Journal, 3*(4), 12–14.
Whitbeck, C. (1996). Ethics as design: doing justice to moral problems. *Hastings Center Report, 26*(3), 9–16.
Working Party (WP) Article 29 D-95/46/EC. (2009). The future of privacy. 02356/09/EN-WP 168.
Yeung, K. (2007). Towards an understanding of regulation by design. In R. Brownsword & K. Yeung (Eds.), *Regulating technologies: Legal futures, regulatory frames and technological fixes* (pp. 79–108). London: Hart Publishing.
Zittrain, J. (2007). Perfect enforcement on tomorrow's internet. In R. Brownsword & K. Yeung (Eds.), *Regulating technologies: Legal futures, regulatory frames and technological fixes* (pp. 125–156). London: Hart Publishing.

Chapter 9
Law as an Artifact: An Assessment

Corrado Roversi

Abstract Artifact theories of law are aimed at explaining the nature of law, that is, answering the ontological question of legal philosophy in terms of the ontology of artifacts in general. In recent years, some legal theorists (particularly Luka Burazin, Jonathan Crowe, Kenneth Ehrenberg, and Corrado Roversi) have been putting forward artifact theories of law to deal with several classic legal-philosophical problems, such as the role of functional explanation in the legal domain, legal normativity, the relation between law and morality, the relation between authoritative production and recognition, and the role of officials in the construction of legal systems. In this article, I will try to highlight and weigh the theory's pros and cons.

Keywords Metaphysics of law · Concept of law · Legal institutions · Artifacts · Social ontology

9.1 Introduction

Recent developments in legal philosophy have shown that the philosophy of technical artifacts and technical functions can have an impact on the question of the nature of law, meaning that an explanation of law can gain a lot from an understanding of law's artifactual nature. While the so-called "ontological problem", as Norberto Bobbio (2011, p. 47ff.) was wont to call it, has always been a part of legal philosophy, several important legal philosophers have been raising significant doubts about

C. Roversi (✉)
Alma Mater Studiorum Università di Bologna, Bologna, Italy
e-mail: corrado.roversi@unibo.it

S. Chiodo, V. Schiaffonati (eds.), *Italian Philosophy of Technology*, Philosophy of Engineering and Technology 35,
https://doi.org/10.1007/978-3-030-54522-2_9

the possibility of addressing the question in a meaningful way. Brian Leiter, for example, has argued that to outline a universal "nature" of legality is a theoretically hopeless enterprise: "If, in the history of philosophy, there is not a single successful analysis of the 'necessary' or 'essential' properties of a human artefact, why should we think law will be different?" (Leiter 2011, pp. 669–670; see also Leiter 2018, Sect. 3). Frederick Schauer (2012) has maintained that it is impossible to single out a universal nature of law encompassing all its possible instantiations in terms of necessary, essential properties: "If concepts can change in this way, and no one appears to have argued that the concepts attached to artifacts are in some mysterious way non-revisable or immutable, then there is no reason to believe that the concept of law cannot change in much the same way" (Schauer 2018, p. 37). Brian Tamanaha has put forward a pluralistic conception under which the idea of a single nature of law should be replaced with that of several different, context-dependent and historically determined structures, each with its own nature: "There is no unique way to categorize social artifacts, so we should not expect there to be a single correct concept or set of characteristics" (Tamanaha 2017, p. 60).

If there is one element these statements clearly share, it is their assumption that law is an artifactual entity, or rather, that its *nature* is artifactual. Hence, it seems that behind all these skeptical conceptions there lies at least one tenet about the nature of law, namely, that law is an artifact, and this is where the philosophy of technology can be brought to bear on the philosophy of law. A few legal philosophers, among whom I include myself as an Italian scholar, have recently been developing artifact theories of law, namely, theories that appeal to the ongoing discussion on the metaphysics of artifacts to explain some features of law, legal systems, and legal institutions. If law is an artifact, what does that mean? Can we derive some interesting consequences from this very broad and generic statement?

Even though not all these theories insist on philosophies of technical artifacts specifically, there is clearly a crucial role that in this discussion is played by problems having to do with functionality, the designer's intentions, and the actual causal mechanisms an artefact needs in order to work (consisting in collective recognition where institutional artifacts are concerned). In this sense, the parallel between law and technology is illuminating in bringing out both the analogies between these two areas of investigation and the tensions. One the one hand, legal philosophy can look to the philosophy of technology for insights about the way in which all artifacts—and not just legal institutions—result from a complex intertwinement of intentions, functional attributions, and selection and repurposing processes. On the other hand, the philosophy of technology can gain from legal philosophy a broader understanding of the way artifactual objects come into existence and of the social factors that play a role in this process. Indeed, some artifacts emerge and are recognized as such ex post, rather than being simply the outcome of authorial creation.

In this article, I will describe the main features of artifact theories of law and try to balance their merits and problems.

9.2 Is There *One* Artifact Theory of Law or a Plurality of Them?

9.2.1 Models in the Metaphysics of Artifacts That Can Be Relevant for the Nature of Law

Philosophical discussion about the metaphysics of artifacts revolves around three concepts: intention, function, and history. Most authors in this field point out that an essential property of artifactual kinds lies in their being *intentionally* created. Risto Hilpinen (1993), in particular, traces the existence of an artifact, as well as its belonging to a given artifactual kind, to an author's intention to produce something of that kind and to an author's accepting that the final result is that kind of thing. Amie Thomasson argues that artifacts depend on a "controlled process of making … involving imposing a number of intended features on the object" (Thomasson 2007, pp. 58–59), and that these creative intentions must include a "substantive concept" of the thing to be created (Thomasson 2003, p. 600).

Other authors underscore in particular the functional nature of the features that are intentionally imposed by an author on her/his creation. In the work of Peter McLaughlin (2001, Chap. 3) and Karen Neander (1991), the functionality of artifacts is explained in terms of their makers' and users' intentions. More specifically, Lynne Baker argues that artifacts have an intended proper function as their primary-kind property, and this makes it possible to distinguish between an artifact and the material aggregate it is made up of (Baker 2004), thus showing that it is possible for artifacts to have an independent metaphysical standing (Baker 2008).

Functional features have also been used to explain the peculiar metaphysics of artifacts within a historico-evolutionary framework: drawing on Ruth Millikan's concept of a direct proper function, meaning the function in virtue of which instances of a given kind are reproduced in the context of an evolutionary history (Millikan 1984. pp. 27–28), Beth Preston has outlined a non-intentionalist view on which an artifact is such in virtue of its having a history of selection or reproduction serving a specific purpose (Preston 1998, p. 243ff.; Preston 2009, p. 226). Along similar lines, Crawford L. Elder analyses artifactual kinds as a specific instance of "copied kinds", namely, kinds of things (be they biological systems or artifacts) that are selected and reproduced to serve a specific proper function by reason of their shape and historical placement (Elder 2007, pp. 38–39): this, in Elder's view, applies as well to cultural artifacts, like etiquette and other kinds of social norms (Elder 2014, p. 40ff.).

Neither the intentions of makers, nor the function their artifacts are meant to fulfil, nor their historical development can on their own explain the metaphysics of artifactual kinds. An objection typically raised against a pure "intention model" is that artifacts built with a certain intention can be used for purposes that differ significantly from what was originally intended—they can even take on purely accidental functions (artifacts are "multiply utilizable", in Preston's sense: see Preston 2009, pp. 227–228)—and another typical objection is that, on this model, any kind of technical construct can be an artifact simply in virtue of its having been intended by an author, even if this author's intentions have no actual support (Houkes and Vermaas 2010, p. 51). On the other hand, a pure "function model" finds an obvious counterexample in all those artifacts that have been built with no clear function in mind (a typical case being works of art, at least on a specific interpretation of them: see Thomasson 2007, p. 57); and if function is considered a primary-kind or essential property, as in Baker's theory, then in cases in which an artifact is repurposed over the course of its history, we should wind up with a different kind, and hence with two coincident objects of two different kinds (Houkes and Vermaas 2010, pp. 146–147); moreover, even an artifact's intended function may not suffice to determine its membership in a given artifactual kind, because to this end it may be necessary to look at other factors, too, such as its shape and structural features (see Thomasson 2007, pp. 57–58, along the lines of Bloom 1996). Finally, a purely historico-evolutionary theory will be hard put to it to explain artifacts of new design and prototypes lacking any evolutionary history (Houkes and Vermaas 2010, pp. 63–64, and a possible reply in Preston 2013, pp. 164–177), and it may also prove to be circular, for it is not clear how you can identify an artifactual kind by referring to the proper functions of its instances, and hence to the previous instances whence comes its reproduction, if you have no substantive concept of the kind that makes it possible to identify such previous instances (see Thomasson 2009, p. 205).

This intertwining of counterarguments shows that for a satisfactory account of artifactual kinds we need to draw on elements extracted from all the three models, and it also shows why these models are almost never taken in their "pure" form but instead combine in various ways. Both Hilpinen and Thomasson, for example, underscore that the creative intentions in play must be successful (Hilpinen 1993, p. 160; Thomasson 2003, p. 598): this brings an element of objective functionality to their theories, but without ruling out the possibility of malfunctioning or borderline artifacts (the success condition posited in their theories is qualified by stating that intention needs to be "largely" successful). Hilpinen (1993, p. 161) concedes that, in most cases, function is a crucial conceptual element of artifactual kinds, though Thomasson insists not on "intended *functions*" but on "intended *features*" broadly conceived (Thomasson 2007, p. 58, italics added). Thomasson, moreover, states that "artifactual kinds are notoriously historical and malleable in nature" (Thomasson 2007, p. 62). Baker insists that her account needs both intentions and function, so much so that artifacts, on her view, are essentially intention-dependent objects (see Baker 2009, p. 83). Preston does not underestimate the role of intentions, either, though she lays emphasis on the *collective* intentions behind the *reproduction* of an artifact rather than on the *individual* intentions behind its original production (Preston 2003, p. 611; 2009, p. 231), and it seems that, in a sense, Elder places the same focus on reproduction within a context, as if the construction of prototypes from scratch were to be considered a borderline case for artifacts (Elder 2007, pp. 39–40).

In the literature, then, we can find models that explicitly merge features from the intention, function, and history conceptions. One such model specifically focuses on technical artifacts: Wybo Houkes and Pieter Vermaas put forward an intention-causal-evolutionary (ICE) theory of functions on which technical artifacts are the outcome of design processes based on a belief that the object produced will lead to the desired goal if properly manipulated according to a use plan (Houkes and Vermaas 2010, Chap. 4). In this model, although design depends crucially on the designers' intention, this intention (and this is where the model is not purely intentional) must be based on a supposedly real causal mechanism: the belief that the use plan will in fact achieve the desired goal must be warranted, and this leads to an evolutionary history of reproduction based on a communicative chain in which the warranted ascription of function and the relevant use plan are transmitted (*ibid.*, pp. 80–84). While Houkes and Vermaas's mixed model is meant to account for technical artifacts, and hence is particularly suited to a functionalist approach, Randall Dipert's historico-intentional model is meant to also explain artifacts whose function is not as clear, particularly in works of art. In Dipert's view, artifacts are intentionally modified objects whose features explicitly communicate their artifactual nature (this in contrast to tools, or instruments, which do not have this communicative feature: Dipert 1995, p. 127); and useful, technical artifacts are only one type of artifact (the other types being the communicative, the expressive, and the artistic: see Dipert 1993, pp. 102–107). Hence, the property of being an artifact is historical: it amounts to having a deliberative history that traces back to plans, meaning that a hierarchical system of interrelated intentions has been put into place by its authors so that other cognitive agents who interact with the artifact will form certain beliefs which they will act on (*ibid.*, p. 54ff., Chap. 7). In Dipert's view, attributing a deliberative history to an artifact is also the best way to understand its functional properties. Thus he considers his historico-intentional conception to be necessary as well to a non-naïve functionalist approach (*ibid.*, p. 91ff.).

An immediate and preliminary difficulty must be addressed if these models are to usefully contribute to an artifact theory of law. Which is to say that, whereas these models propose to explain the metaphysics of *material* objects, law, legal systems, and legal institutions are inherently immaterial: they mainly consist of representations, and their physical substratum, when present, is relevant only insofar as it supports these representations. Moreover, these representations are inherently social, and hence collective, whereas material objects can be exclusively individual: as Thomasson (2007, p. 52) notes, "[u]nlike social and institutional objects, the existence of artifacts doesn't seem to presuppose any collective intentions of any kind – it makes perfect sense to suppose that a solitary human could create a knife, though not a government or money". So the question comes up, how can a philosophy of material culture contribute something important to a domain of abstract, inherently social entities?

That artifacts may be abstract is a possibility that does find discussion in the philosophical literature. Dipert, for example, contrasts the abstractness of novels with the more "sensuous" qualities of other kinds of works of art, and he claims that abstract artifacts "typically involve more conscious interpretation" (Dipert 1993. p. 163), which is certainly an interesting point that could apply to law as well. And Hilpinen, for his part, qualifies literary works as types, setting them in distinction to all the copies that can be made of them, these copies being tokens (Hilpinen 1993, p. 171ff.). "In the case of music and literary arts", he explains, his theory leads "to a view of the identity of works of art which resembles David Lewis's conception of works of fiction, according to which a fiction should be identified by an act of storytelling and not just by the text produced by such an act" (*ibid.*, 173). Elsewhere, Hilpinen clearly states that "[o]ntologically, an artifact can be a concrete particular object such as the Eiffel Tower, a type (a type object) which has or can have many instances (for example, a paper clip or Nikolai Gogol's *Dead Souls*), an instance of a type (a particular paper clip), or an abstract object, for example, an artificial language" (Hilpinen 2004). But it is certainly Thomasson's theory of fictions that deals most clearly with the case of immaterial artifacts. Thomasson sets out an artifactual theory of fictions (Thomasson 1999, pp. 13–14, Chap. 3) proceeding from two kinds of existential dependence: *historical* and *constant* dependence. As artifacts, fictions and fictional characters are *historically* dependent (for their existence) on an author's intentions, but in virtue of their abstract nature they are also *constantly* dependent on a given text.

Thomasson discusses as well what she calls "public" artifacts, in which connection she also deals with the *social* nature of artifacts. Drawing on Dipert's idea of artifacts as objects specifically conceived to communicate their nature, Thomasson clarifies that some artifacts are explicitly built with some receptive features intended to signal that they belong to a given kind, so that the specific audience to which this membership is signalled can interact with such artifacts in an appropriate way, that is, in accordance with an appropriate set of shared social norms (Thomasson 2014, pp. 49–51). Examples of this kind are churches and flags. Such public artifacts, as Thomasson explains them, blur the distinction between artifactual and institutional kinds, precisely because some of their intended features depend on social norms, and social norms depend not simply on the intentions of the *makers* of artifacts but rather on the *shared* intentions that are formed within a group (*ibid.*, p. 55).

With these models and qualifications in mind, we can now turn to artifact theories of *law*.

9.2.2 Differences and Common Elements Among Artifact Theories of Law

Artifact theories of law were developed in close temporal proximity (from 2014 to 2018), and in large part independently, by four authors: Luka Burazin, Jonathan Crowe, Kenneth Ehrenberg, and myself.[1]

[1] It is important to note, however, that several authors have previously worked on the analogy between legal institutions and artifacts, though they did not systematically develop the analogy by bringing to bear the philosophy of technology and artifactual kinds. Thus, for example, Bruno Celano (1999, pp. 249–258) attributes to Hans Kelsen the view that legal norms are "intentional artifacts" (*artefatti intenzionali*), namely, meanings (*contenuti di senso*) based on a creative,

Burazin (2016, 2018) adopts an intention model, drawing substantially on Hilpinen and Thomasson. In his theory, artifacts are used not to explain legal *institutions*—an idea he qualifies as "not new" (Burazin 2016, p. 385)—but to explain law as a legal *system*. Legal systems are conceived by him as "abstract institutional artifacts", combining the intention-authorial model of artifacts with a socio-ontological conception framed along the lines of Searle and Thomasson. On this approach, artifacts are objects that, in order to come into being as objects of a given kind, depend on a largely successful authorial intention (*ibid.*, pp. 388–389): a screwdriver depends on a designer's intention. Institutions, by contrast, consist of constitutive rules that set forth existence conditions for tokens of institutional kinds and are based on collective acceptance (*ibid.*, pp. 393–394): I can be said to be standing in line at the post office only if the line of people I am in correctly instantiates the shared rules about what a line of people at the post office is. Hence, institutional artifacts are institutional objects whose existence conditions in constitutive rules require an author (*ibid.*, p. 395). In the case of legal systems, the relevant authors are the legal officials, and what constitutes the system is their shared concept of validity, of primary and secondary rules, and of the overall constitutional framework (*ibid.*, p. 398), while collective acceptance of the fundamental constitutive rules defining legal authority is rooted in the social community (*ibid.*, p. 397). What we have here, then, is a two-layered ontological structure of developed legal systems: a social *norm* of recognition coupled with a *rule* of recognition, the former embraced by the members of the *community* (perhaps tacitly, but necessarily in the we-mode; Burazin 2018, pp. 114–119), the latter shared among *officials* (*ibid.*, p. 120). Burazin holds that not all systems are developed and structured in this way. It can also happen that, at an initial stage of development, the social norm of recognition simply unifies a set of social norms into a body constituting the legal system (*ibid.*), in which case the authors of the legal systems are the members of the community themselves (Burazin 2016, p. 396). This genealogical dimension is meant to capture some features of the historico-evolutionary model for explaining artifacts. Indeed, drawing on Thomasson, Burazin concedes that the authorial concept behind artifacts can change diachronically, and hence that in the mind of officials the concept of a legal system is "susceptible to change" (*ibid.*, p. 398). Finally, Burazin applies to legal systems the success condition that Hilpinen and Thomasson introduce to address the problem of support in creating technical artifacts, to which end he translates this condition into a requirement of social validation: a legal system as conceived by officials must correspond to an actual practice of norm-following and of applying sanctions in case of noncompliance (Burazin 2018, pp. 129–134).

---

authorial act of lawmaking and whose efficacy is of general scope, a view that in several respects resembles the one developed in full-fledged artifact theories of law. Castelfranchi and Tummolini (2006) have also put forward a theory of institutions in general as coordination artifacts, namely, "artifacts such that the recognition of their use by an agent and the set of cognitive opportunities and constraints (deontic mediators) are necessary and sufficient conditions to enable a multiagent coordinated action" (*ibid.*, p. 320).

Crowe (2014), too, previously espoused an intention account of artifacts drawn mainly from Hilpinen. What he theorizes, however, is not a two-*layered* ontology, like Burazin's, but a two-*sourced* one, putting forward an intention-acceptance theory of artifact kinds. Artifacts, and institutional artifacts in particular, can be the outcome of an authorial intention, *or* they can be the outcome of collective acceptance (Crowe 2014, p. 747), either case requiring a successful realization (*ibid.*, p. 748ff.), meaning that the artifact cannot come into being absent that intention or acceptance. It is a crucial role that this "success condition" plays in Crowe's view (indeed, the condition is drawn here, too, from the intention model, where its purpose is to avert the problem of actual support in fulfilling a technical function). Given that, in Crowe's conception, the features that explain artifacts are mainly functional, if law (as an artifact) is to be non-defective, it must be able to actually perform its function, which is typically that of serving as a "deontic marker", meaning that law "marks the boundaries of permissible social conduct" (*ibid.*, p. 751). This functionalist approach leads Crowe to bring in some elements of a pure function model of artifacts in support of a natural-law conception. In his view, law must be able to create a sense of social obligation, and its "metaphysics of malfunction" (Baker 2009) therefore entails that, if law fails to include a core moral content, it will be defective (Crowe 2014, pp. 754–755).

Like Crowe, Ehrenberg (2014, 2016, 2018) also moulds an artifact theory of law within a functionalist framework, but unlike Crowe he does not inject any natural law into this framework but rather insists on the *function* of law as a crucial part of any legal-positivistic explanation of it.[2] A crucial element of Ehrenberg's theory lies in the *desiderata* that Preston posits for any theory of artifacts (see Preston 2009, p. 214ff.; Ehrenberg 2016, p. 120ff.), and in particular in the idea that artifact functions must be "multiply realizable", that artifacts must be "multiply utilizable", and that they can malfunction. In Ehrenberg's view, law is a specific genre of social institutions conceived as "abstract institutionalized artifacts": "It is the broadest institution whose function it is to generate and/or validate institutions" (*ibid.*, pp. 12, 191). There is no function peculiar to law: law is not a "functional kind" in the sense that anything fulfilling that particular function is necessarily law (*ibid.*, pp. 50, 76–77), because several other kinds of social institutions can perform the most general function of law (*ibid.*, 121–122), namely, creating or modifying normative reasons for action for the members of a given social community. Moreover, and coherently with Preston's historico-evolutionary model, Ehrenberg argues that the functions of legal institutions can vary and change, and even be "phantasmatic", and legal institution can malfunction in several ways, to the point that they can even be defective in creating normative "deontic powers" when they run out of recognition within a community (*ibid.*, 119–128). Here, again, it is crucial to refer to Searle's model of institutional facts: the institutionality of law makes it possible to clarify

[2] A functionalist account along legal-positivistic lines has also been developed by Himma (2018), who argues that legal systems as artifacts have the conceptual function of regulating behaviour as a means of keeping the peace, and they do so by means of authorized coercive-enforcement mechanisms.

how abstract legal artifacts can have a degree of persistence and continuity even without any actual recognition (*ibid.*, p. 108ff.), whereas the artifactuality of law explains a crucial aspect of its normativity. Just as Dipert shows that artifacts typically "communicate" their artifactual nature, and Thomasson clarifies how "public artifacts" serve a specific recognitional function, Ehrenberg holds that a typical function of law (a macro-function) is to signal that some institutions are in fact legal, so as to make it explicit that political authority intends these institutions to be valid in the most general way and to shape and reframe the community members' reasons for action (Ehrenberg 2014, p. 263ff.; 2016, pp. 137–139, 175ff.; 2018, p. 182ff.).

This insistence on phenomena of repurposing, modification, and diachronic evolution also figures centrally in my own "historico-intentional" model of the artifactuality of legal institutions (Roversi 2016, 2018). This model takes up Dipert's concept of "deliberative history" by tracing artifactuality to a historical property rooted in an original "creative process" consisting of authorial intentions and in a series of further modification, reinterpretation, and development processes (Roversi 2016, p. 218ff.; 2018, pp. 95–99): legal institutions are therefore, on this view, the outcome not just of an original authorial intention but also, and more significantly, of a *history* of intentions. Moreover, my model makes use of a concept in Houkes and Vermaas's intentional-causal-evolutionary theory of technical artifacts: the concept of a "use plan". In Houkes and Vermaas's theory, artifacts are connected with use plans, understood as ordered, goal-directed sets of actions that must be carried out using the object in question in order for it to fulfil its function (Houkes and Vermaas 2010, p. 18ff.). Similarly, in my view, legal institutions as immaterial artifacts have an interaction plan in conditional form (if X, then normative consequence Y follows) specified by constitutive rules. Legal institutions are thus immaterial rule-based artifacts (Roversi 2016, pp. 224–245, 226–240). Moreover, just as Houkes and Vermaas argue that it must be possible to ascribe some causal support to artifactual functions, I make the argument that a legal institution's constitutive rules and the related interaction plans can work only if based on collective acceptance as a general mechanism. Here, again, the argument is crucially predicated on Searle's theory of social institutions. Differently from Crowe, and similarly to Burazin, I do not assume that an artifact theory needs to be connected with any specific theory of law's functionality: legal institutions can serve different purposes. But, similarly to Crowe, I stress that the artifactuality of legal institutions makes them reason-based objects (*ibid.*, pp. 225–226, 230–231): they are subject to evaluation on the basis of technical and teleological rationality. Moreover, the historico-intentional nature of legal institutions as artifacts is such that, in my view, their nature and content may not be entirely transparent to us or immediately fixed by actual intentional states in the legal community (Roversi 2018, pp. 103–105). The view I take here in regard to legal institutions as artifacts is similar to what in Elder's aforementioned historico-evolutionary model is argued about artifacts broadly and cultural artifacts more specifically.

From this brief presentation we should be able to appreciate in what respects the artifact theories of law so far developed differ and what they have in common. They are certainly different, to an extent, in their object: Burazin's focus is on legal systems, Crowe's and Ehrenberg is on law as a genre of legal institutions, mine is on legal institutions broadly. This partial difference in focus, however, does not imply that the relative theories cannot be combined. Indeed, they can be conceived as more or less specific ways to describe law's artifactuality: the artifact "law," in Crowe's and Ehrenberg's sense, makes it possible to qualify legal systems in Burazin's sense and legal institutions in my sense as *legal* artifacts. Further, Crowe's and Ehrenberg's specific insistence on law's functionality does not stand in contradiction to Burazin's model or mine, both of which take a quite neutral view on this point. Indeed, Ehrenberg's view of jurisprudents as "modest joiners"—meaning that they are open to the possibility that functional attributions to legal institutions can vary and be qualified as well on the basis of empirical and historical research (Ehrenberg 2016, p. 139ff.)—seems perfectly compatible with my and Burazin's neutral stance on functionalism. The most idiosyncratic result of the insistence on law's functionality seems to be Crowe's claim that law's artifactuality can lead to a sort of rationalistic natural-law view, particularly because neither my model nor Burazin's or Ehrenberg's is predicated on a strong cognitivist conception of morality. But to some extent this gap can be filled, to which end it should be pointed out that (1) Crowe embraces a context-dependent cognitivist view, on which "natural law is objective and normative, but nonetheless socially embodied, historically extended and dependent on contingent facts about human nature" (Crowe 2019: introduction); (2) Ehrenberg acknowledges that legal institutions are meant to have a distinctive general effect on reasons for action; and (3) my own model connects the defectiveness of legal artifacts with considerations of (technical or teleological) rationality.

There are clearly several common elements to these theories, to the point that they can form the basis on which to put forward a unified artifact theory of law. This unified theory is one I would describe as follows. Law is a genre of abstract institutional artifacts, in the sense that, within a social community, it is collectively recognized to confer a specific status on other institutions, signalling that these institutions are meant by political authority to hold generally and to shape the community members' normative reasons for action. Legal institutions are thus created as institutional, abstract artifacts in this framework: they are rule-based artifacts built to enable or foreclose interaction among human agents. In *informal* contexts, they are simply the outcome of a creative process rooted in human intentions—an outcome that, at some point in the development of this process, is collectively recognized as such, namely, as having created an institution. Like a tree that becomes a bench, certain regular activities become normative, and the institution that develops with them begins to be referred to as "that institution" and as "legal". In *formal* contexts, collective acceptance supports constitutive rules defining the conditions of legal authority and hence the status of legal officials. Legal institutional artifacts thus become the outcome of the creative intentions of legal officials. The most general legal artifact in this sense is a legal system, meaning a set of legal institutional artifacts defining the conditions under which legal norms are valid and may be produced and applied. On the basis of these conditions of validity, several other institutional sub-artifacts can be created. In informal and formal contexts alike, legal artifacts need collective acceptance as the mechanism enabling them to actually work, that is, to accordingly shape the community members' normative reasons for action. In all cases, in other words, the makers' intentions must find a significant degree of success, that is, it must be that they are actually practiced and enforced. To this end legal artifacts must integrate considerations of rationality in two forms: technical rationality, serving as a basis on which to determine whether a legal artifact is properly framed to achieve its purpose, and teleological rationality, on which basis to determine whether this purpose can find meaningful support in the community. These considerations of rationality are very much context-sensitive, so the approach they require is a "modest" one, open to empirical research, particularly when it comes to defining the purpose of a given legal institution. For example, what purpose universities serve, and under what conditions academic structures are considered to be well-suited to that purpose, very much depends on the social and axiological context in which these considerations are made. Thus, reasoning from purely utilitarian premises, universities can be said to primarily serve the purpose of training people for the skills they need to enter the job market, and in this sense they can be understood to be inevitably subject to market pressures; from a more deontological point of view, however, their purpose can be assessed in quite a different, almost contrary way: it is to advance knowledge and support the independent research and teaching which makes that possible, *regardless* of market forces.

Let us now consider some explanatory advantages of this unified artifact theory of law, as well as some of its problems.

9.3 What Does an Artifact Theory of Law Explain?

9.3.1 The Advantages

Having laid out the main features of an artifact theory of law, we know that it can be framed as a unified theory. Given such a unified theory, however, why should we use it to work out a metaphysics of law in the first place? What are its explanatory advantages?

The first such advantage relates to the normativity of law. It is a central concern of legal philosophy to explain how law can provide normative reasons for action, especially so when working within a legal-positivistic paradigm, where it is critical that a theory be able to explain whether, and if so how, law can have a normative force that cannot be reduced to that of morality or that of merely prudential considerations. This independent normative force has been compellingly explained by Ehrenberg (2016, Chap. 7), who does so by embedding an artifact theory in a functionalist framework, showing that law as an institutional artifact can gain its independent normative force in virtue of both its institutionality and its artifactuality. To be sure, Ehrenberg (*ibid.*, Chap. 7, Sect. E) explains law's normativity by drawing on Searle's theory of desire-independent reasons for action, thus implying that an explanation of law's normativity in terms of its institutionality is not a merit of an artifact theory *per se*. However, the theory does have the merit of pointing out that law's normative force can depend in part on the normativity of recognition typical of normative artifacts, as Ehrenberg (2014, p. 263ff.; 2016, Chap. 7, Sect. F; 2018, p. 182ff.) does in line with Dipert and Thomasson. Moreover, as Auke Pols (2013) shows through Dancy's model of practical reasoning, artifacts can indeed provide reasons for action by prescribing or enabling action. Hence, it seems that an attempt can be made to trace legal normativity to the general, inherent normativity of artifacts.

Despite these achievements, however, it is doubtful that the normativity of artifacts can fully explain the normativity of law and of social institutions. On the contrary, there are reasons to think that the converse is true, namely, that the normativity of *artifacts* has a social dimension. As Marcel Scheele has shown, the definition of an artifact's "proper use" often depends on the formation of "opinion groups", to the effect that "social features are necessary conditions for justified ascription of proper functions" (Scheele 2006, p. 32). Houkes and Vermaas similarly show how an artifact's use plans become normative on the basis of a process of testimony (between designers and users, and among users in a community) that reflects a process of "social entrenchment" and "institutionalization" (Houkes and Vermaas 2010, p. 114). Moreover, where artifacts are concerned, normativity always depends on a technical, hypothetical structure, and there is reason to doubt that the same holds necessarily for law as well. It could instead be argued that law can provide us with robust, normative, and context-independent reasons for action (a discussion of this question can be found in Ehrenberg 2016, Chap. 7, Sect. C). So the contribution an artifact theory of law can give in answering the problem of law's normativity can at best be partial.

In my own work (Roversi 2018, p. 99ff.), I argue that an advantage of an artifact theory of law is that it is well-suited to accounting for a typical dialectic in legal theory: that between the legal authorities' *production* of law—the core element of law in traditional formalistic legal positivism—and the community's *recognition* of law, as well as the *enforcement* through which the law becomes effective in the community, which interest is instead central to legal realism. This dialectic could be understood in light of the double existential dependence of immaterial, abstract artifacts (as discussed in Thomasson 1999, Chaps. 2 and 3), meaning that, on the one hand, these artifacts are *historically* dependent for their existence on an original, authorial creative process, but at the same time their continued existence is also *constantly* dependent on a collective recognition lacking which they would expire. In light of that dialectic, we can work together two of the core tenets of legal realism and legal positivism.[3] Central to legal realism is the insight that norms are shaped and modified in the practice of their application, so much so that concepts like "legislative intent" or the "true meaning of a provision" can be thought to be

[3] Here I understand legal realism mainly along the lines of Scandinavian legal realism, and so as a general theory about the nature of law and not as a contingent description of legal adjudication, as Leiter (2007, Chap. 1) instead interprets American legal realism.

illusory. So interpreted, an artifact can be explained only in light of (a) its *user function*, namely, the function it comes to have within a certain community of users, and (b) its dependence on its gaining actual recognition within that community. On the other hand, a crucial assumption of formalistic legal positivism is that norms and legal institutions find their content in the original authorial act of the legislator and in the latter's creative intention: so interpreted, an artifact can be explained only in light of (a) its *design function*, namely, the function as imagined by its original designer, and (b) its historical dependence on an original act of creation. So, by drawing a distinction between user and design function, and another one between historical and constant dependence, an artifact theory of law can show that both formalistic legal positivism and rule-sceptic legal realism are reductionist accounts, in that they capture only one relevant aspect of the phenomenon they intend to explain.[4]

The intertwining of the concepts of function, authorial intention, and history that lies at the core of the metaphysics of artifacts can also illuminate some aspects of legal interpretation—and here I would count a third advantage of an artifact theory of law. In legal interpretation a distinction is traditionally drawn between originalist interpretation—based on the idea of the legislator's original intent or the provision's original linguistic meaning—and dynamic, evolutionary (or also teleological) interpretation, which underscores the need to adapt provisions to new circumstances the legislator either did not or could not foresee (see for example Guastini 2011, pp. 100–101). If we look at this distinction from the point of view of an artifact theory of law, we can see it as a particular application of a more general distinction made in the interpretation of artifacts. Dipert (1993, p. 87), for example, points out "two possible goals for the interpretation of artifacts …: 1. Historicism or antiquarianism. To determine the historical genesis and history of an artifact: that is, finding out, as best one can, an object's deliberative history. 2. Functionalism. To determine the most useful function an object can now be conceived as having". Of course there is an important peculiarity of legal interpretation. As Dipert points out, when we interpret artifacts, "the pursuit of the function of an object – how we should conceive of and use it – should … always be our chief aim", and understanding an artifact's history is useful only because this "sometimes is the best way to achieve an understanding of its function" (Dipert 1993, pp. 90–91). This does not necessarily hold for legal artifacts. Here, evolutionary arguments are not inevitably more relevant than originalist ones, and indeed the contrary can often be true. That is because legal artifacts are authoritative: they trace their origin to political authority, and as a result the original function envisaged by the designer carries a special, authoritative status compared to the normal status of an artifact's author. But the intertwining of those interpretive arguments in the context of law is simply an instance of interpreting artifacts generally conceived. A full-fledged theory of legal interpretation along similar lines is developed by Crowe (2019, Chaps. 11–12), drawing on the interpretation of artifacts generally conceived.

[4]A similar point is argued by Bruno Celano (1999, p. 254), explaining Kelsen's "principle of effectiveness" in terms of "intentional artifacts."

9.3.2 The Challenges

The just-mentioned peculiarity in the interpretation of legal artifacts—namely, that legal artifacts are authoritative in a way other artifacts are not—brings out an important limit of an artifact theory of law, a limit that can be expressed in a succinct formula: authors are not authorities. Even though, as Houkes and Vermaas (2010, p. 110ff.) show, designers of artifacts can have a special status in attesting to an artifact's use plan, the status of political authority with regard to legal artifacts cannot be subsumed under that of designers. Legislative intent is not binding simply because of a special epistemic access that legislators have to the content of the provisions they enact: what makes such intent binding, rather, is that legislation is supposed to express an exercise of legitimate political power. In a sense, the relation here is reversed: with artifacts, we become authors in virtue of our having *made* the object, but where provisions are concerned, we need to first have authority (the authority conferred on us within a community)—*only then* can we proceed to enact the provision we intend to enact. Hence, there is a crucial dimension of law—the problem of the legitimacy of legal authority—that an artifact theory cannot adequately capture in terms of the author-user relation. As Crowe (2019, Chap. 10) argues, an option we can pursue in order to attack this problem would be to shift the source of legitimacy from authority to social coordination.

There is another significant challenge confronting an artifact theory of law. As Burazin (2016, p. 399) has aptly stated, in conceiving of law as an artifact, we seem to be making the assumption that all legal institutions have authors and that these authors can be clearly identified, but of course institutions have an extensive evolution, and very often they emerge within a community as the outcome of a process that lacks explicit authors and can very well be tacit. How can an artifact theory of law explain this customary nature of the origin of legal institutions? This is a crucial problem, and one that is not confined to the customary origins of law. In taking up this problem, for example, Dan Priel argues that an account of law in terms of artifacts and design intentions is inconsistent with common law practice: in his view, lurking behind the idea that legal institutions are the outcome of an explicit design is a rationalistic ideology according to which law is "a tool designed by humans to ultimately attain a better life", which means "a more moral life" (Priel 2018, p. 259). The common law, on the other hand, is framed according to a traditional view under which "law … is not put in place where custom fails; it is a continuation of, a manifestation of, custom" (*ibid.*, p. 264).

Customs and custom-like legal practices are a problem that has been addressed by all scholars who have engaged with the artifact theory. Burazin (*ibid.*, p. 396) speaks of "informal institutional artifacts", having the community as a whole as their original author, and indeed he concedes that the artifact theory of law seems "to adopt a very broad concept of authorship … The artifact theory of law does not preclude collective authorship and accepts as authors a wide range of persons, including those who sustain the artifact in question and its active users" (*ibid.*, p. 399). Similarly, as mentioned, Crowe includes collective acceptance among the possible creators of an artifact, precisely to account for those customary-like phenomena in which (as in the case of a tree becoming a bench) artifacts emerge as kinds of socially created objects: these he calls "unintentionally created artifacts" (Crowe 2014, p. 746). Ehrenberg clarifies that customs, at a certain point, come to be based on a conscious decision to "raise the level of the customary rule to the point of seeing it appropriately officially or communally enforced" (Ehrenberg 2016, p. 123), which amounts to something similar to Burazin and Crowe's "collective authorship" phenomenon. In a similar vein, my model introduces the concept of "recurrence" artifacts, namely, artifacts created by recurrent action "whose content is in some sense related to the artifact but is not an intention to create an artifact of that kind" (Roversi 2016, p. 221).

Now, to be sure, these solutions require to a certain degree that the concept of an artifact be "stretched"—something that some may not be ready to accept. Thomasson, for example, explicitly excludes terms like *path* and *village* from the class of artifactual kinds, limiting her discussion to what she calls "essential artifactual" terms, understood as "terms that necessarily have in their extension all and only artifacts, considered as such (as intended products of human action)" (Thomasson 2003, p. 593). Hilpinen (1993, p. 167ff.), on the other hand, explicitly accepts the existence of collective artifacts, and in this class includes (well-designed) towns, where the parts have been produced "in such a way that each part fits its surroundings architecturally and aesthetically, and contributes to the 'wholeness' of the town".[5] However, such a stretching of the concept of artifacts is something that other authors call for in the legal domain even in an extreme way: Leiter (2018, p. 11), for example, suggests that "we think of the category of 'artifacts' more broadly, as the category of phenomena that *result from human action*, which are *responsive to human interests*, and which are not otherwise natural kinds". Indeed, that legal artifacts require a broader conception of artifacts may well be not so much a shortcoming of those theories as a significant improvement of artefact theories of law over theories of other kinds of artifacts, and of technical artifacts in particular. This in two ways. First, one is not forced to assume a priority of material, authored artifacts in explaining artifacts in general. And, second, the general observation that artifacts can be the outcome of emergence, rather than of explicit creation, can help us understand several instances of material artifacts as well. Not all material artifacts are like screwdrivers. Some are like Gothic churches: they can be incomplete, defective, even contradictory in their design plans, and may have taken a long time to become the way they are now. This is something a general theory of artifacts needs to be able to explain, and here general theory can gain a lot by drawing on a theory of legal artifacts in particular.

The problem of customs and emerging institutions, and its tension with the concept of artifacts, brings into light another major challenge that an artifact theory of law must face. All these theorists agree that legal artifacts are socially rooted; that

[5] It is important to note that these collective artifacts, in Hilpinen's view, collect artifacts having their own authors, whereas the collectively created artifacts envisaged by artifact theorists of law are single artifacts created through an act of recognition which is collective, and which may even extend over time.

they very often simply depend on collective recognition, which is in itself creative; and that in this sense they are "institutionalized" or "institutional". It wouldn't be unreasonable, therefore, to ask why we should prefer an artifact theory of law over a socio-ontological account relying on the idea of institutions based on collective acceptance and a standing practice. There is one thing that needs to be clarified, however, before we take up this question. Which is to say that, in a sense, an artifact theory of law *already is* a socio-ontological account. There is no frontal opposition between these two accounts, because legal institutions, conceived as artifacts, are described by all the artifact theorists of law as based on collective recognition and acceptance. An artifact theory of law does not claim to *replace* a socio-ontological, institutional theory of law, but rather proposes to *specify* such a theory. So the question, more accurately framed, is not why an artifact theory of law should be preferred over a socio-ontological account—for if the theory is justified, so is its account of law—but rather whether the added element an artifact theory of law brings to a socio-ontological account is really needed.

Here is my answer. There is a crucial thing that users and designers of artifacts share and that is not necessarily part of a purely socio-ontological account. Which is to say that they both deal with objects, entities. An artifact theory of law focuses on the peculiar way in which the legal domain is a domain of abstract objects, and not simply of facts about people behaving in a certain way and believing certain things. It explains how law can be made up of hypostatized, symbolic things, and how these things can organize normative behaviour. It might be replied that we need not explain this "objectual" aspect of law, or that this aspect is parasitic or irrelevant. But that is not the case. It has been argued, for example, that from an evolutionary standpoint the characteristic capacity of human beings to attribute legal status and build hierarchies of competence has evolved from the same development of the brain that made it possible to build objects endowed with symbolic value, like ornaments, or with sacred ritual meaning, as in the case of religious burials (Dubreuil 2010, Chap. 3). In the construction of symbolic artifacts and the construction of legal institutions we have two aspects of the same, distinctly human ability. Law is not simply a matter of socially shared norms: it consists of normative structures charged with symbolic value, and to ignore this metaphysical feature of the legal domain would be tantamount to dismissing an important part of its genealogy. An artifact theory of law, then, enhances the explanatory capacity of a pure socio-ontological account by showing how law is embedded within the broader, distinctly human practice of building artifactual objects (material or immaterial) endowed with symbolic meaning. But, of course, we need the socio-ontological background to explain the shared, inherently collective nature of these objects.

9.4 Conclusion

In this article, I have set out to answer some questions about the meaning and significance of an artifact theory of law. I began by showing how, even if the different artifact theories so far advanced depend on different models of the metaphysics of artifacts, it is possible to merge these theories within a single, unified account. In a word, an artifact theory of law explains law as made up of abstract, immaterial artifacts of a specific kind, in that they require us to interact on the basis of constitutive rules grounded in collective acceptance, making such artifacts in this sense institutionalized.

I then underscored the main advantages of this account, while also pointing out the main challenges it faces. An artifact theory of law can explain several aspects of the reason-giving nature of legal artifacts, as well as the dialectic between the production and recognition of law that lies at the core of the debate between legal positivism and legal realism and behind the tension between originalist and evolutionary legal interpretation. However, insisting on the artifactual character of legal institutions could be taken to entail that the crucial moment for these institutions' existence lies in their creation, and consequently that all legal institutions need an author—a conclusion immediately falsified by the original, customary nature of law. Moreover, the author's "authoritative" status over the use of an artifact is much weaker than the normative legitimate status of political authority: artifactual authorship cannot explain legal authority. I have tried to show that customs and emerging institutions are not necessarily a problem for an artifact theory, but I conceded that the concept of authorship assumed by a theory of this kind is broader than is usually assumed, and that the question of legitimacy falls outside its scope.

Finally, I have argued that an artifact theory of law is not meant to replace a socio-ontological account but rather to improve on it, and that its contribution consists, among other things, in explaining the "objectual" aspect of legal institutions, namely, the nature of legal structures as symbolic, normative objects—a characteristic owed to their genealogy and evolution and grounded in the capacities of modern human beings.

References

Baker, L. R. (2004). The ontology of artifacts. *Philosophical Explorations, 7*, 99–111.

Baker, L. R. (2008). The shrinking difference between artifacts and natural objects. *American Philosophical Association Newsletter on Philosophy and Computers, 7*, 2–5.

Baker, L. R. (2009). The metaphysics of malfunction. *Techné, 13*, 82–92.

Bloom, P. (1996). Intention, History, and Artifact Concepts. *Cognition, 60*, 1–29.

Bobbio, N. (2011). *Giusnaturalismo e positivismo giuridico*. Roma-Bari: Laterza. (1st ed. 1965).

Burazin, L. (2016). Can there be an artifact theory of law? *Ratio Juris, 29*, 385–401.

Burazin, L. (2018). Legal systems as abstract institutional artifacts. In L. Burazin, K. E. Himma, & C. Roversi (Eds.), *Law as an artifact* (pp. 112–135). Oxford: Oxford University Press.

Castelfranchi, C., & Tummolini, L. (2006). The cognitive and behavioral mediation of institutions: towards an account of institutional actions. *Cognitive Systems Research, 7*, 307–323.

Celano, B. (1999). *La teoria del diritto di Hans Kelsen*. Bologna: il Mulino.

Crowe, J. (2014). Law as an artifact kind. *Monash University Law Review, 40*, 737–757.

Crowe, J. (2019). *Natural law and the nature of law*. Cambridge: Cambridge University Press.

Dipert, R. R. (1993). *Artifacts, art works, and agency*. Philadelphia: Temple University Press.

Dipert, R. R. (1995). Some issues in the theory of artifacts: defining artifacts and related notions. *The Monist, 78*, 119–135.
Dubreuil, B. (2010). *Human evolution and the origins of hierarchies: The state of nature*. Cambridge: Cambridge University Press.
Ehrenberg, K. M. (2014). Law's artifactual nature: How legal institutions generate normativity. In G. Pavlakos & V. Rodriguez-Blanco (Eds.), *Reasons and intentions in law and practical agency* (pp. 247–266). Cambridge: Cambridge University Press.
Ehrenberg, K. M. (2016). *The functions of law*. Oxford: Oxford University Press.
Ehrenberg, K. M. (2018). Law is an institution, artifact, and practice. In L. Burazin, K. E. Himma, & C. Roversi (Eds.), *Law as an artifact* (pp. 177–191). Oxford: Oxford University Press.
Elder, C. L. (2007). On the place of artifacts in ontology. In E. Margolis & S. Laurence (Eds.), *Creations of the mind: Theories of artifacts and their representation* (pp. 33–51). Oxford: Oxford University Press.
Elder, C. L. (2014). Artifacts and mind-independence. In M. Franssen, P. Kroes, T. A. C. Reydon, & P. E. Vermaas (Eds.), *Artefact kinds: Ontology and the human-made world* (pp. 27–43). Berlin: Springer.
Guastini, R. (2011). *Interpretare e argomentare*. Milano: Giuffrè.
Hilpinen, R. (1993). Authors and artifacts. *Proceedings of the Aristotelian Society, 93*, 155–178.
Hilpinen, R. (2004). Artifact. In: *Stanford encyclopedia of philosophy*. Winter 2011 edition. https://plato.stanford.edu/archives/win2011/entries/artifact/
Himma, K. E. (2018). The conceptual function of law: Law, coercion, and keeping the peace. In L. Burazin, K. E. Himma, & C. Roversi (Eds.), *Law as an artifact* (pp. 136–159). Oxford: Oxford University Press.
Houkes, W., & Vermaas, P. E. (2010). *Technical functions: On the use and design of artefacts*. Berlin: Springer.
Leiter, B. (2007). *Naturalizing Jurisprudence: Essays on American Legal Realism and Naturalism in Legal Philosophy*. Oxford: Oxford University Press.
Leiter, B. (2011). The demarcation problem in jurisprudence: A new case for scepticism. *Oxford Journal of Legal Studies, 31*, 663–677.
Leiter, B. (2018). Legal positivism about the artifact law: A retrospective assessment. In L. Burazin, K. E. Himma, & C. Roversi (Eds.), *Law as an artifact* (pp. 3–28). Oxford: Oxford University Press.
McLaughlin, P. (2001). *What functions explain: Functional explanations and self-reproducing systems*. Cambridge: Cambridge University Press.
Millikan, R. G. (1984). *Language, thought, and other biological categories*. Cambridge: MIT Press.
Neander, K. (1991). The teleological notion of "function". *Australasian Journal of Philosophy, 69*, 454–468.
Pols, A. (2013). How artefacts influence our action. *Ethical Theory and Moral Practice, 16*, 575–587.
Preston, B. (1998). Why is a wing like a spoon? A pluralist theory of function. *The Journal of Philosophy, 95*, 215–254.
Preston, B. (2003). Of marigold beer: A reply to Vermaas and Houkes. *The British Journal for Philosophy of Science, 54*, 601–612.
Preston, B. (2009). Philosophical theories of artifact function. In A. Meijers (Ed.), *Philosophy of technology and engineering sciences* (pp. 213–233). Amsterdam: Elsevier/North Holland.
Preston, B. (2013). *A philosophy of material culture: Action, function, and mind*. New York and London: Routledge.
Priel, D. (2018). Not all law is an artifact: Jurisprudence meets the common law. In L. Burazin, K. E. Himma, & C. Roversi (Eds.), *Law as an artifact* (pp. 239–267). Oxford: Oxford University Press.
Roversi, C. (2016). Legal metaphoric artifacts. In J. Stelmach, B. Brozek, & L. Kurek (Eds.), *The emergence of normative orders* (pp. 215–280). Kraków: Copernicus Center Press.

Roversi, C. (2018). On the artifactual – and natural – character of legal institutions. In L. Burazin, K. E. Himma, & C. Roversi (Eds.), *Law as an artifact* (pp. 89–111). Oxford: Oxford University Press.

Schauer, F. (2012). On the nature of the nature of law. *Archiv für Rechts- und Sozialphilosophie, 98*, 457–467.

Schauer, F. (2018). Law as a malleable artifact. In L. Burazin, K. E. Himma, & C. Roversi (Eds.), *Law as an artifact* (pp. 29–43). Oxford: Oxford University Press.

Scheele, M. (2006). Function and use of technical artefacts: social conditions of function ascription. *Studies in History and Philosophy of Science, 37*, 23–36.

Tamanaha, B. (2017). *A realistic theory of law*. Cambridge: Cambridge University Press.

Thomasson, A. (1999). *Fiction and metaphysics*. Cambridge: Cambridge University Press.

Thomasson, A. (2003). Realism and human kinds. *Philosophy and Phenomenological Research, 67*, 580–609.

Thomasson, A. (2007). Artifacts and human concepts. In E. Margolis & S. Laurence (Eds.), *Creations of the mind: Theories of artifacts and their representation* (pp. 52–73). Oxford: Oxford University Press.

Thomasson, A. (2009). Artifacts in metaphysics. In A. Meijers (Ed.), *Philosophy of technology and engineering sciences* (pp. 191–212). Amsterdam: Elsevier/North Holland.

Thomasson, A. (2014). Public artifacts, intentions, and norms. In M. Franssen, P. Kroes, T. A. C. Reydon, & P. E. Vermaas (Eds.), *Artefact kinds: Ontology and the human-made world* (pp. 45–83). Berlin: Springer.

Chapter 10
Technical Normativity

Alberto Artosi

Abstract Despite their increasingly expanding empire, technical norms have scarcely attracted the attention of legal scholars. This article tries to understand the working of such norms by approaching the issue from the standpoint of the notion of "artefactual normativity"—i.e. normativity concerning artefacts and their characteristics—and then by analysing a strikingly growing class of technical norms more or less explicitly directed at influencing behaviour. What emerges from this analysis is a non-legal form of normativity which involves the subject in a process of continuous observation, surveillance and direction of his/her conduct whose final aim is his/her self-identification with certain imposed standards. The paper ends with a section on the standardisation of the language of regulation.

Keywords Technical norms · Artefactual normativity · Indicators · Legal norms · Standardization

> Instrumental action is governed by *technical rules* based on empirical knowledge.... [I]nstrumental action organizes means that are appropriate or inappropriate according to criteria of an effective control of reality... *Incompetent* behavior, which violates valid technical rules or strategies, is condemned per se to failure through lack of success; the "punishment" is built, so to speak, into its rebuff by reality (Habermas 1968, pp. 91–92).

> A finalized system thus gradually arises in which the logic of performance threatens to prevail over the inherent desirability of the objective pursued (Ost and van de Kerchove 1987, p. 524).

A. Artosi (✉)
Alma Mater Studiorum Università di Bologna, Bologna, Italy
e-mail: alberto.artosi@unibo.it

S. Chiodo, V. Schiaffonati (eds.), *Italian Philosophy of Technology*, Philosophy of Engineering and Technology 35,
https://doi.org/10.1007/978-3-030-54522-2_10

10.1 Introduction

Despite their increasingly expanding empire, technical norms have scarcely attracted the attention of legal scholars. This may be due to several reasons, first of all their uncertain normative status which makes them look like a sort of "unidentified normative object", as Benoit Frydman puts it in his extensive study of the argument (Frydman 2014, p. 9). Indeed, in the absence of an agreed meaning of "technical norm", technical norms are usually defined by reference to their form as a document. For example, the official ISO/IEC definition of technical standard—another term for technical norm—is

> a document, established by consensus and approved by a recognized body, that provides, for common and repeated use, rules, guidelines or characteristics for activities or their results, aimed at the achievement of the optimum degree of order in a given context.[1]

As "voluntary agreements" based on inclusive consensual processes,[2] technical norms are not mandatory, nor do they constitute a fixed and immutable legislation, "but undergo periodic review to ensure that they take account of the latest technological developments and market trends". This sensitivity to technological progress entails another crucial feature of technical norms: they "codify the latest technology and… are therefore an invaluable source of knowledge… they distil expert knowledge and make it available to all". The conclusion one can draw comes hardly as a surprise to those familiar with the classical philosophical account of technical norms as normative/cognitive hybrids.[3] But this does not help very much if our aim is to understand the working of such norms. A more promising approach is suggested by Vaesen (2013). We turn to this approach in Sect. 10.2.

---

[1] ISO and IEC stand, respectively, for the International Organization for Standardization and its "sister organization", the International Electrotechnical Commission. The definition of a standard and the following quotations in the text are found at the ISO/COPOLCO site (https://www.iso.org/sites/ConsumersStandards/1_standards.html).

[2] The ISO/IEC Guide (see the preceding note) defines consensus as the result of a systematic procedure for removing opposition and dissent: "consensus is general agreement, characterized by the absence of sustained opposition to substantial issues by any important part of the concerned interests and by a process that involves seeking to take into account the views of all parties concerned and to reconcile any conflicting arguments. … Consensus is an essential procedural principle, requiring the resolution of substantial objections. … The aim is to resolve substantive issues before the final stages of development". Once these final stages are reached, any conflict and discussion about the standards adopted is banished to the advantage of "technical harmonization". On consensus see also Sect. 10.5, notes 7 and 8 and the text corresponding to the notes.

[3] See, e.g., Azzoni (1991).

10.2 Artefactual Norms

The gist of Vaesen's (2013) approach is that, far from considering technical norms as a normative arcane, it accounts for a specific form of technical normativity he calls "artefactual normativity". On this account, "The difference between agent and artefactual normativity would be related to what norms are about: in agent normativity, norms are about human beings and their actions, artefactual normativity on the other hand concerns artefacts" and their characteristics (Vaesen 2013, p. 121). Artefactual norms occur usually in evaluative processes in which the goodness of an artefact is being assessed in terms of its compliance with some institutionally established standards. Of those norms some pertain to the artefact's function, and others concern other—non-functional—characteristics the artefact needs to have in order "to be able to compete the multitude of other artefacts on the market and to be able to safeguard the other values and goals we cherish" (Vaesen 2013, p. 134). Vaesen distinguishes three kinds of non-functional norms (Vaesen 2013, pp. 123–127):

- Norms establishing standards of compatibility between different artefacts, e.g. the WC3 (World Wide Web Consortium) technical standards for web communication technologies. A compatibility norm specifies how an artefact A should be made in order to be used in connection with other artefacts A_1, A_2,…, A_n.
- Norms regulating "the extent to which an artefact may affect other entities – people, other artefacts, or the environment" (Vaesen 2013, p. 124), e.g. norms establishing safety standards such as the BRC (British Retail Consortium) Global Standard for Food Safety or the ISO 14001 Environmental Management System.
- Norms specifying the characteristics that an artefact should have in order to ensure that its quality conforms to certain standards of goodness—such as effectiveness, reliability, usability, aesthetic appeal, cost-effectiveness, etc.—to meet its users' needs and expectations, e.g. the IMQ (Italian Institute of the Mark of Quality) quality standards for technological artefacts.

This tripartite classification needs not be taken too literally as the three kinds of norms often overlap. For example, environmental and safety standards are included in the IMQ quality standards. But this is not the problem. What matters is that such kinds of norms, though representative of a large class of technical norms, do not exhaust the domain of technical regulation. A strikingly growing body of technical norms produced and used in contemporary governance processes are in fact more or less explicitly directed at influencing behaviour. These norms are revealing of a kind of technical normativity that evidently falls outside the scope of Vaesen's approach in terms of artefactual (i.e. non-agent) normativity. In what follows I shall try to shed some light on this kind of normativity by reference to a class of technical norms that play an ever more prominent role as regulatory devices at a global level.

10.3 Indicators

Although the term "indicator" has become of current use to refer to such peculiar class of technical norms, it is endowed with a variety of meanings that makes it very difficult to provide a fairly uncontroversial definition. In my discussion, therefore, I shall rely heavily on Davis et al.'s (2012) reconstruction of the concept.[4] According to their definition,

> An indicator is a *named* collection of *rank-ordered* data that purports to represent the past or projected performance of different units. The data are generated through a process that *simplifies* the data about a complex social phenomenon. The data, in this simplified and processed form, are capable of being used to *compare* particular units of analysis (such as countries or institutions or corporations), synchronically over time, and to *evaluate* their performance by reference to one or more standards (Davis et al. 2012, p. 6).

I have italicized some key words to emphasise the salient features of indicators: naming, ordering, simplification, comparison and evaluation.

Naming an indicator does not only identify the phenomenon for which it purports to provide a representation—e.g. "human development"—but it may even create it, as is the case with the GPA (Grade Point Average) Index used to measure academic performance.

- Indicators aim at simplifying complex phenomena by reducing them to numerical representations to enhance comparison with other numerically represented phenomena. The entailed process is highly selective: simplification is achieved by filtering and excluding "certain data, including outliers or other data deemed to be unreliable or irrelevant" (Davis et al. 2012, pp. 6–7).
- Indicators set evaluation standards, e.g. the indicators used by the QS World University Rankings set the standards against which the performance of each university is to be measured in order to determine its place in the ranking.

It is immediately evident that indicators share many features of artefactual norms (in fact, both kinds of technical-normative devices fall under the ISO definition of a standard, respectively under "activities" and "results"). In particular:

- Both artefactual norms and indicators are standard-setting tools: they set technical requirements that have to be met in order to improve goodness and/or performance.
- In both cases the standards are derived from and codify expert knowledge.
- Both artefactual norms and indicators are used in evaluative processes.
- Both kinds of norms aim at improving efficiency, reliability and cost-effectiveness of products and decision-making processes.
- Both artefactual norms and indicators have regulatory effects.

[4] For other attempted but substantially convergent definitions see Boisson De Chazournes (2014, pp. 69–70) and Restrepo Amariles (2014, p. 232).

What, then, is the difference between the two kinds of technical norms? Not unexpectedly, the answer is that the difference is "related to what norms are about": unlike artefactual norms, indicators are about agents, both collective agents—such as states, institutions, organisations, etc.—and individual agents, such as university professors. This immediately raises the problem of the relationship between the kind or normativity embedded in indicators and other forms of agent—in particular legal—normativity. To tackle this problem, I will first take a closer look at the way indicators operate as behaviour-influencing devices.

10.4 Indicators as Behaviour-Influencing Tools

A lucid characterisation of the way indicators affect behaviour is found in Büthe (2012). In this paper we read:

> Research in psychology and behavioral economics has shown that knowing being observed leads human beings to change their behavior in ways that ... the observed expects to elicit greatest approval from the observer. And indicators communicate the desired behavior (or more precisely what the observer would like to see) subtly but very clearly through scores and rankings, making them ideal instruments for eliciting behavioral changes through observation (Büthe 2012, pp. 43–44).

Since this remark about the modus operandi of indicators evokes Bentham's Panopticon, I shall refer to it as "panopticity". Indicators are "panoptical" exactly in the sense that they influence the observed's behaviour in the desired direction by putting him under the pressure of a continuous surveillance by an omnipresent observer. Panopticity clearly implies that the observer has decided in advance what he/she "would like to see" so as to elicit exactly—and confer by that very fact predictability to—those behavioural changes he wishes to observe. Secondly,

> The ultimate targets of an indicator face both material and socio-political incentives for behavioral adaptation – probably, though not necessarily, toward the ideal defined by the indicator (Büthe 2012, p. 44).

This remark clarifies three further aspects of the way indicators make the observed's behaviour change in the direction of the observer's desired objective: they act as regulative ideals bringing behavioural adaptation through enforcement mechanisms constituted by various kinds of—direct and indirect—incentives. These incentives operate quite independently of their being agreed upon by the observed. For example,

> Law school deans and faculty may mostly be opposed to the use of the indicator as a measure of the quality of legal education, but they overwhelmingly feel compelled to pay close attention and seek to improve their school's rank, given that prospective students and future employers of those potential students are using this indicator (Büthe 2012, p. 44).

In this sense, the adoption of an indicator by some agents becomes in itself an incentive to other agents to adapt their own behaviour to the ideal defined by the indicator. Finally,

> Indicators – and the resulting rating or rankings – should make easier for interested parties within the social group or even beyond…to bring pressure on the targets to "improve" their behaviour in such a way that it yields a higher score (i.e., a highest rating or maybe even a higher ranking) (Büthe 2012, p. 45).

That is to say that indicators act as regulative ideals in the proper sense that they not only operate as tools of behavioural adaptation, but compel the observed to tend toward the ideal defined by the indicator in a continuous effort to improve adaptation. Frydman (2014) has something to add to this analysis of the indicators' normativity. He identifies a peculiar mechanism through which an indicator operates in order to secure the observed's compliance with the (observer's) desired objective:

> The management by objectives occurs through one of the most powerful and interesting technologies of the new management: benchmarking. Its very simple principle consists in fixing the level of performance to be achieved by reference to a model defined in general by the performance of another company that is a market leader (Frydman 2014, p. 30, my translation, here and below).

The benchmarking works

> on the basis of a principle of enhancing performance by x% over the previous year, in a process of continuous improvement of quality, return, profit, etc. What is more, the technology of management by indicators and of benchmarking may be specified in order to analyse, within the company itself, performance over time of the agents and of the teams and to compare them with each other.… The technology of benchmarking makes it possible to specify management by objectives at the level of each service and each agent. It also makes it possible to rank performance in both areas and to publish the ranking in order to praise and reward the top performers and blame and penalize the stragglers (Frydman 2014, pp. 30–31).

The result is a regime of permanent evaluation which, thanks to suitable positive and negative incentives, stimulates interpersonal but, more importantly, intrapersonal competitiveness:

> These devices, coupled with appropriate incentives (rewards or penalties, promotions or dismissals), create powerful motivations, but also strong pressures, among workers in the company and economic agents in general, who react strongly and often even anticipate them by adapting their behaviour to them. The device is at the same time effective and inexpensive to the extent that the agents themselves become their own supervisors (Frydman 2014, p. 31).

Or, to put it in psychological terms, they are lead to "interiorize"[5] the device, thus becoming the omnipresent observers of themselves.

---

[5] See Vinokur (2014, p. 236) concerning the devices of "normalization" of the university. The formation of "interiorized" habits is explicitly designated as the objective of the new forms of "flexible" regulation by Valaguzza (2017, p. 121).

10.5 What Kind of Normativity?

We started by raising the question of the relationship between normative standards aimed at influencing behaviour and other forms of agent normativity. We have identified several features of the way in which those standards operate. In conclusion, we should at least be able to understand that we are faced with a distinctive *modus operandi*: more than being subject to change *through norms*, the agent's behaviour is subject to a continued process of *evaluation of its conformity* to the standards. As a consequence, agents are not only induced to conform to the standard, but, what is more, they are led to lose sight of the very possibility of not conforming. What emerges is a form of normativity other than, not to say opposed to (and even in competition with), the kind of normativity involved in the view that law is addressed to autonomous, self-directing agents who take norms as *guides* for their conduct.[6] This raises a lot of questions, especially the question of legitimacy (Frydman 2014, pp. 54–59). The argument advanced by all current definitions of technical norms is that they draw their legitimacy from consensus.[7] Unfortunately, consensus is a notoriously suspect criterion, and even more suspect when it is the result of procedures that have it not only as their object but also as their goal.[8] Even worse, the emphasis on consensus betrays the underlying ideological framework, for as a source of legitimation consensus falls outside the scope of traditional, i.e. democratic, law-making procedures:

> Modern democracy demands not consensus regarding political programs and goals, but acceptance of an institutional framework which on the one hand guarantees peaceful competition between these divergent programs and goals, and on the other provides a procedure which – despite rivalry and divergence – enables political decisions to be made. This procedure can be circumscribed with the terms "election" and "majority". Accordingly, governments are not formed through consensus and the legitimacy of laws does not depend upon a *consensus omnium* (Bayertz 1994, p. 46).[9]

[6] Is it an accident that scholars of the new forms of regulations speak of *influencing* behaviour? E.g., Davis et al. (2012, pp. 10, 15); Fisher (2012, p. 237); Dadush (2012, p. 407), and most paradigmatically Sunstein (2016). Now, the aim of legal norms is not to *influence* behaviour. According to a reasonably reliable and agreed characterisation, a legal norm entails at least the following features: (1) it is a binding rule of conduct issued by the competent state authority; (2) it determines right and duties of the subjects to whom it is addressed; and (3) abidance by subjects is enforced by the state coercion. Such a norm does not violate the subjects' autonomy. On the contrary, it presupposes it.

[7] See note 2 above and the text corresponding to the note.

[8] As the ISO/IEC Guide emphasises, consensus is not synonymous with unanimity. This is correct. In fact, according to the ISO procedure it is up to the president of the competent expert committee to decide if and when there is sufficient consensus about the proposed standards (Van Waeyenbege 2014, p. 99).

[9] Of course, as it is well known, it is not at all difficult to abstract the "institutional framework" of democracy from its political content. For example, the ISO procedures for the elaboration of standards contemplates "majority" (Van Waeyenbege 2014, pp. 96–97; Carreira Da Cruz 2014, p. 128), but that doesn't mean that they are *democratic*. As a matter of fact, "the exchanges between stakeholders are based upon scientific, technical or empirical data. This is reflected into the fact that the

To be sure, indicators, like all other kinds of technical norms, can—and often do—acquire legal validity by being given a legislative basis. But the mere fact of being included in the state's legislation does not make them become *law* in the sense previously explained that they should provide normative guidance to their subjects. Indeed, their normative effects tend to be more like those of an interiorised form of control: the subjects are brought to conform their behaviour to—and to measure their "value" against—standards that they have introjected. No wonder that, in a society slaved to a logic of influence aimed at making each individual the censor of himself, the controls are being introjected. But, as Herbert Marcuse pointed out over 50 years ago in his classical analysis of the forms of social control prevailing in the advanced industrial society,

> the term "introjection" perhaps no longer describes the way in which the individual by himself reproduces and perpetuates the external controls exercised by his society. Introjection suggests a variety of relatively spontaneous processes by which a Self (Ego) transposes the "outer" into the "inner". Thus introjection implies the existence of an inner dimension distinguished from and even antagonistic to the external exigencies (Marcuse 1964, p. 12).

Standardisation involves the subject in a process of continuous observation, surveillance and direction of his conduct whose final aim is his self-identification with the imposed standards. "In this process, the 'inner' dimension of the mind in which opposition to the status quo can take root is whittled down. The loss of this dimension, in which the power of negative thinking – the critical power of Reason – is at home" (Marcuse 1964, p. 13), is enforced by the presumption that the standards are derived from objective, scientifically grounded facts, with the consequence that non-compliant, i.e. non-performant, behaviour is perceived not only as irrational but even as "rebuffed" by reality.[10] The upshot is a blind positivity that makes any dissent from the standards both ideologically and practically infeasible in the face of a plain surrender to the established facts. On the other hand, the reduction of the normative-evaluative to the cognitive-descriptive dimension brought about by the standards marks the twilight of the conception of the law as grounded in the dualism between fact and norm, *Sein* and *Sollen*, intellect and will. But this reduction of the normative to the factual is only possible because the "facts" from which technical norms are alleged to flow—the "facts" that these norms are assumed to be a description of—are already laden with normativity, not to say that they are constituted by the norms themselves. To quote again from Marcuse, "If the individuals

participants are appointed to the different committees because of their expertise in the field concerned and not by a political legitimation" (Van Waeyenbege 2014, p. 99, my translation). This shift from democratic legitimation to a legitimation based on technical expertise has far reaching depoliticising effects. As Lisa Duggan remarks in his *Twilight of Equality?*, "The most successful ruse of neoliberal dominance is the definition of economic [more in general public] policy as primarily a matter of neutral, technical expertise. This expertise is then presented as separated from politics and culture, and not properly subject to specifically political accountability or cultural critique" (Duggan 2003, p. xiv).

[10] See the quotation from Habermas at the beginning of the paper.

find themselves in the things which shape their life, they do so, not by giving, but by accepting the law of things—not the law of physics but the law of their society" (Marcuse 1964, p. 13).

10.6 From Coercion to Terror

The replacement of the normativity of law by a normativity of performance has been variously predicted in the last century, e.g. by Luhmann (1969), Lyotard (1979) and Ost and van de Kerchove (1987). In particular, Lyotard has put the advent of a "logic of maximum performance" at the heart of his diagnosis of the "postmodern condition". According to Lyotard, the "mastering" of reality allowed by technology is the source of a self-legitimating power—"self-legitimating, in the same way a system organized around performance maximization seem to be" (Lyotard 1979, p. 47)—able

> to guide individual aspirations through a process of "quasi-apprenticeship", "free of all disturbances", in order to make them compatible with the system's decisions. The decisions do not have to respect individuals' aspirations; the aspirations have to aspire to the decisions, or at least to their effects. Administrative procedures should make individuals "want" what the system needs in order to perform well (Lyotard 1979, p. 62).

Unlike the law, this process needs not operate through coercion as it entails in itself "a certain level of terror, whether soft or hard: be operative (that is commensurable) or disappear" (Lyotard 1979, p. xxiv). In our terminology here, this means that any deviation from the standards is punished not by penalty but by the "exclusion from the game".[11] In this sense, terror appears as the ultimate improving-efficiency tool: it is

> the efficiency gained by eliminating, or threatening to eliminate, a player from the language game one shares with him. He is silenced or consents, not because he has been refuted, but because his ability to participate has been threatened. ... The decision makers' arrogance ... consists in the exercise of terror. It says: "Adapt your aspirations to our ends – or else" (Lyotard 1979, p. 63).

As Karl Kraus once wrote: "One of the most widespread diseases is diagnosis". The problem is that the diagnoses of the last century have more and more become descriptions of our present and enjoy today the authority of the realised prophecies.[12]

---

[11] For example, the university professors who do not achieve the levels of "efficiency" prescribed by the indicators of the Italian National Agency for the Evaluation of Universities and Research Institutes (ANVUR) are excluded from participating in the national commissions for the National Scientific Qualification (ASA). Nor is this the only "exclusion from the game". For a penetrating criticism of the evaluation procedures of the Italian National Agency see Pinto (2012).

[12] Thanks to Andrea Gentili for having suggested this thought to me.

10.7 Note on the Standardisation of Language

Standardisation aims at simplification and homogenisation. These processes are mirrored in language. As standardisation progresses, language increasingly expresses itself in standardised forms that can be easily associated with predetermined modes of thought and behaviour. The style—just read a standardisation document—flattens itself into an abridged, condensed, simplified syntax. Language closes up, becomes ritualised. The ritualised language deploys its power in the act of naming. Naming becomes a "creative" act (see Sect. 10.3 above). At the same time, language degenerates into a vehicle for conveying formulaic, and hence indisputable, truths. Such ritualisation of language finds its extreme form in the overwhelming use of abbreviations. Acronyms dominate the linguistic scene of standardisation. The List of Abbreviation of Davis et al. (2012) contains 127 abbreviations, from ACER (Australian Council for Educational Research) to WHO (World Health Organisation)—and these are only a sample of the huge number of current abbreviations. To be sure, use of abbreviations is justified by practical needs. But there is something more. As Marcuse remarked long ago,

> The abbreviations denote that and only that which is institutionalized in such a way that the transcending connotation is cut off. The meaning is fixed, doctored, loaded. Once it has become an official vocable, constantly repeated in general usage, "sanctioned" by the intellectuals [i.e. by the experts], it has lost all cognitive value and serves merely for recognition of an unquestionable fact (Marcuse 1964, pp. 97–98).

In other words, abbreviations contribute to making language a radically depoliticised and dehistoricised technical instrument at the service of the dominant behavioural rationality. The eradication of history from language has here an enormous significance. For not only words have a history, but history, and through it culture and society, are encoded into their very meaning. So, the eradication of history from language is at the same time the eradication of history from culture and society, with the result that society is doomed to an uncritical acceptance of the present. In fact, history is the realm of the negative: it is the locus of all contradictions, illusions, aspirations, ambiguities and bewilderments, hopes and deceptions, of all actualised as well as unactualised possibilities; briefly: it is the realm of *facts*—not the ordered, sterilised, unambiguous "facts" identified by the indicators,[13] but facts endowed with a genuine, i.e. historical, reality. The suppression of history means the substitution of such realm of facts with a host of pale and mystifying ghosts.

Acknowledgements I wish to thank Silvia Vida for her inspiring assistance during the writing of this paper. I owe a great deal to her and Matteo Galletti's book (Galletti and Vida 2018).

[13] "Indicators… often present the world in black and white, with few ambiguous intermediate shades. They take flawed and incomplete data that may have been collected for other purposes, and merge them together to produce an apparently coherent and complete picture" (Davis et al. 2012, pp. 8–9).

References

Azzoni, G. M. (1991). *Cognitivo e normativo: il paradosso delle regole tecniche*. Milano: Angeli.

Bayertz, K. (1994). The concept of moral consensus: Philosophical reflections. In K. Bayertz, *The concept of moral consensus: The case of technological interventions in human reproduction* (pp. 41–57). Dordrecht: Springer.

Boisson De Chazournes, L. (2014). Standards, régulation internationale et organisations internationales. In B. Frydman & A. van Waeyenberge (Eds.), *Gouverner par le standards et les indicateurs: De Hume aux rankings* (pp. 69–92). Brussels: Bruylant.

Büthe, T. (2012). Beyond supply and demand: A political-economic conceptual model. In K. E. Davis, A. Fisher, B. Kingsbury, & S. Engle Merry (Eds.), *Governance by indicators: Global power through quantification and rankings* (pp. 29–51). Oxford: Oxford University Press.

Carreira Da Cruz, M.-A. (2014). La responsabilité sociétal comme champ d'extension de la norme: le cas d'ISO 26000. In B. Frydman & A. van Waeyenberge (Eds.), *Gouverner par le standards et les indicateurs: De Hume aux rankings* (pp. 117–144). Brussels: Bruylant.

Dadush, S. (2012). Impact investment indicators: A critical assessment. In K. E. Davis, A. Fisher, B. Kingsbury, & S. Engle Merry (Eds.), *Governance by indicators: Global power through quantification and rankings* (pp. 392–434). Oxford: Oxford University Press.

Davis, K. E., Kingsbury, B., & Engle Merry, S. (2012). Introduction: Global governance by indicators. In K. E. Davis, A. Fisher, B. Kingsbury, & S. Engle Merry (Eds.), *Governance by indicators: Global power through quantification and rankings* (pp. 3–28). Oxford: Oxford University Press.

Duggan, L. (2003). *The twilight of equality? Neoliberalism, cultural politics, and the attack on democracy*. Boston: Beacon Press.

Fisher, A. (2012). From diagnosing under-immunization to evaluating health care systems: Immunization coverage indicators as a technology of global governance. In K. E. Davis, A. Fisher, B. Kingsbury, & S. Engle Merry (Eds.), *Governance by indicators: Global power through quantification and rankings* (pp. 217–246). Oxford: Oxford University Press.

Frydman, B. (2014). Prendre le standards et les indicateurs au sérieux. In B. Frydman & A. van Waeyenberge (Eds.), *Gouverner par le standards et les indicateurs: De Hume aux rankings* (pp. 5–65). Brussels: Bruylant.

Galletti, M., & Vida, S. (2018). *Libertà vigilata. Una critica del paternalismo libertario*. Roma: IF Press.

Habermas, J. (1968). Technology and science as ideology. In J. Habermas, *Towards a rational society: Student protest, science, and politics* (pp. 81–122). Cambridge: Polity Press.

Luhmann, N. (1969). *Legitimation durch Verfahren*. Neuwied: Luchterhand.

Lyotard, J.-F. (1979). *The postmodern condition: A report on knowledge*. Manchester: Manchester University Press, 1984.

Marcuse, H. (1964). *One-dimensional man: Studies in the ideology of advanced industrial society* (2nd ed.). London: Routledge, 1991 (reprinted 2007).

Ost, F., & van de Kerchove, M. (1987). *Jalons pout une théorie critique du droit*. Brussels: Publications des Facultés universitaires Saint-Louis.

Pinto, V. (2012). *Valutare e punire*. Napoli: Cronopio (reprinted 2014).

Restrepo Amariles, D. (2014). The mathematical turn: L'indicateur Rule of Law dans la politique de développement de la Banque Mondiale. In B. Frydman & A. van Waeyenberge (Eds.), *Gouverner par le standards et les indicateurs: De Hume aux rankings* (pp. 193–234). Brussels: Bruylant.

Sunstein, C. R. (2016). *The ethics of influence: Government in the age of behavioral science*. Cambridge: Cambridge University Press.

Vaesen, K. (2013). Artefactual norms. In M. J. de Vries, S. O. Hansson, & A. W. M. Meijers (Eds.), *Norms in technology* (pp. 119–136). Dordrecht: Springer.

Valaguzza, S. (2017). Nudging pubblico vs. pubblico: nuovi strumenti per la regolazione flessibile di ANAC. *Rivista della regolazione dei mercati*, 1, 91–121.

Van Waeyenbege, A. (2014). Le norms ISO, CEN et celles issues des consortiums privés: bric à brac ou système pour l'Union europeénne? In B. Frydman & A. van Waeyenberge (Eds.), *Gouverner par le standards et les indicateurs: De Hume aux rankings* (pp. 93–115). Brussel: Bruylant.
Vinokur, A. (2014). La normalization de l'université. In B. Frydman & A. van Waeyenberge (Eds.), *Gouverner par le standards et les indicateurs: De Hume aux rankings* (pp. 235–261). Brussels: Bruylant.

Part III
Philosophy of Science

Chapter 11
Research Programs Based on Machine Intelligence Games

Guglielmo Tamburrini and Francesco Altiero

Abstract Games have played a significant role throughout the history of artificial intelligence and robotics. Machine intelligence games are examined here from a methodological perspective, focusing on their role as generators of research programs. These research programs are schematized in terms of framework building, subgoaling, and outcome appraisal processes. The latter process is found to involve a rather intricate system of rewards and penalties, which take into account the double allegiance of participating scientists, trading and sharing interchanges which occur in multidisciplinary research environments, in addition to expected industrial payoffs, and a variety of other research benefits in the way of outreach and dissemination of results, recruitment of junior researchers and students' enrolment. Examples used here to illustrate these various aspects of the outcome appraisal process include RoboCup and computer chess, Go, Poker and video-games. On the whole, a reflection on research programs that are based on machine game playing opens a window on central features of the complex systems of rewards and penalties that come into play to appraise machine intelligence investigations.

Keywords Technological research programs · Methodology of AI · Methodology of robotics · Research gamification · Multiple research allegiances

G. Tamburrini (✉) · F. Altiero
Università degli Studi di Napoli Federico II, Naples, Italy
e-mail: guglielmo.tamburrini@unina.it; francesco.altiero86@gmail.com

S. Chiodo, V. Schiaffonati (eds.), *Italian Philosophy of Technology*, Philosophy of Engineering and Technology 35,
https://doi.org/10.1007/978-3-030-54522-2_11

11.1 Introduction

Games have played a significant role throughout the history of artificial intelligence (AI from now on) and robotics.[1] Pioneers of AI—including Alan Turing, Claude Shannon, John McCarthy, Allen Newell and Herbert Simon—identified chess as a fruitful domain for AI inquiry and contributed to early investigations in computer chess (Cordeschi 2002, pp. 177–179). More recently, substantive AI research focused on Poker and Go. The entry "Games and puzzles" figures in the top-level list of AI research topics compiled by the AAAI (Association for the Advancement of Artificial Intelligence; http://www.aaai.org/home.html). RoboCup is an ongoing soccer competition for robotic teams involving research groups worldwide. International funding agencies have made ample provisions for robotic competitions, early paradigmatic examples of which are the DARPA Grand Challenge held in 2004, 2005 and 2007[2] and the DARPA Robotic Challenge, started in December 2013 and ended in June 2015.[3]

A chief motive of methodological interest for games in AI arises from their clear rules of behavior and success criteria. On this account one may address, and often sidestep successfully, familiar conceptual difficulties besetting the definition of other forms of intelligent behavior (Turing 1950; Marr 1977) and hindering their principled empirical investigation. Moreover, games afford suitable task environments to investigate central issues for AI and robotics, such as heuristic search, action planning and coordination, learning and knowledge representation. Competitions based either on AI social games (such as card, board and video games) or on robotic team games additionally provide significant test beds and benchmarks for ranking playing performances of AI and robotics systems, as these emerge from the confluence in one system of the above multiple skills and capabilities. For these reasons, AI and robotic games (collectively referred to as machine intelligence games from now on) have repeatedly played the role of *research programs generators* in AI and robotics.

The special role of machine intelligence games as generators of research programs in AI and robotics is analysed here from a methodological perspective, by reference to the processes of *framework building*, *subgoaling*, and *outcome appraisal*. Framework building and subgoaling are the pillars of major two-process models of scientific inquiry (Godfrey-Smith 2003). These include Thomas Kuhn's model of normal science (Kuhn 1962), Imre Lakatos' model of scientific research programs (Lakatos 1970), and Larry Laudan's model of research traditions (Laudan 1977). More distinctive traits of research programs based on machine intelligence

[1] This contribution builds on (Tamburrini 2015), expanding on the methodological framework introduced there, and considering more recent research directions in machine intelligence games. All URLs mentioned in the text were last accessed on September 7th, 2019. We are grateful to the editors of this volume for their useful comments on a previous version.

[2] https://web.archive.org/web/20120110114231/http://archive.darpa.mil/grandchallenge.

[3] https://archive.darpa.mil/roboticschallenge

games are found in their outcome appraisal process, especially in view of the fact that inquiries on machine intelligence games are *multidisciplinary hubs*: heterogeneous research communities in AI and robotics converge there, with their wide range of competences, methodologies, technological platforms, research interests, goals and expectations. Accordingly, the system of rewards and penalties adopted by each one of these research communities contributes to appraise progress and benefits, if any, flowing from multidisciplinary research programs based on machine intelligence games.

To explore the complexities involved in this outcome appraisal process, we leverage on the notion of double allegiance of scientists participating in research programs based on machine intelligence games. Here, we identify the primary allegiance of an AI or robotics scientist with the well-established research area(s) within which this scientist carries out most of her problem-solving activity (such as artificial perception, machine learning, multi-agent systems, and so on). From their secondary allegiance to machine intelligence games, these scientists expect to gain insights and results that are significant from the standpoint of their primary allegiance. For example, RoboCup raises significant challenges for scientists with a primary research allegiance to multi-agent systems research. Robotic soccer playing is interesting from their primary allegiance standpoint, as it requires one to develop increasingly better algorithmic solutions to cooperative and competitive action planning and coordination problems. And the broader interest of these solutions for multi-agent systems research is often established by their successful transfer to a variety of other domains characterized by tight temporal constraints on action planning and coordination. The "RoboCup Championship and Symposium" was introduced exactly to explore the interest of RoboCup technologies and systems for major research issues in traditional areas of AI and robotics (Visser and Burkhard 2007, p. 128).

The appraisal of machine game inquiries requires one to take duly into account the double allegiance of participating scientists, the regional system of rewards associated to their primary allegiance, and the benefits for their primary allegiance that flow from their secondary allegiance to machine game inquiries. Thus, the notion of double allegiance is useful to bring to the fore chief motivations of scientists who committed themselves to research on machine intelligence games since the early days of AI. Claude Shannon (Shannon 1959) identified computational solutions to chess playing, which are of no practical importance in themselves, as holding solutions to major problems in practically important domains.

Secondary allegiances to research programs based on machine intelligence games wax or wane in the light of appraisal criteria that are based on primary allegiances. Noteworthy from this perspective is the initial involvement and the ensuing withdrawal from computer chess investigations of scientists focusing on human problem solving and its computer simulation: the use of more powerful hardware resources and statistical learning methods, while improving computer chess performances, progressively weakened the secondary allegiance of scientists expecting to gain from chess programs significant insights into the heuristic problem solving strategies of humans and other cognitive systems (Cordeschi 2002, p. 178 and

p. 182; Cordeschi and Tamburrini 2015). Even more important, after the historical victory of Deep Blue against Garry Kasparov,[4] the overall interest of computer scientists in the game of chess declined in favour of Go and other more complex games, insofar as the game of chess held no longer additional promises from their primary allegiance perspectives.

Secondary allegiances to research programs on machine intelligence games are additionally motivated by expectations of fruitful trading and sharing activities occurring within multidisciplinary research environments. Moreover, as Shannon pointed out early on, the integration of machine intelligence technologies which is needed to build game playing machines is expected to feed the pipeline between academic research and industry. Last but not least, the gamification of machine intelligence research is usually expected to bring with it various fringe benefits, by motivating junior researchers, attracting wider cohorts of science and engineering students, and raising public awareness and interest towards AI and robotics. On the whole, machine intelligence games enable one to highlight a rich network of regional interests, rewards and diminishing returns, trading and sharing interactions within academic research groups, as well as between the latter and industrial R&D groups, in addition to ancillary motivational and communication benefits that jointly contribute to determine the waxing and waning of multidisciplinary research programs. To explore these various themes, let us turn to consider first RoboCup's long-term objective of robotic soccer teams competing on equal terms with the best human soccer teams.

11.2 RoboCup's Regulative Goal and Framework Building

When RoboCup started in 1997, the conjecture was initially advanced that a team of fully autonomous robots would eventually defeat the human world champion soccer team by the year 2050. Later versions of the RoboCup manifesto more cautiously suggested that the goal of beating the best human soccer team "will take decades of efforts, if not centuries. It is not feasible, with the current technologies, to accomplish this goal in any near term" (http://www.robocup.org/objective/). Since robotic soccer teams are expected to perform poorly against human teams for many years to come, RoboCup's ultimate goal is neither a realistic objective for the current generation of robotic engineers to pursue nor a useful benchmark to assess progress in robotic soccer playing from one RoboCup tournament to the next. Therefore, one may legitimately wonder why this elusive target was proposed at all, and whether it has any effective role to play in the context of current robotics research.

The RoboCup manifesto addresses these concerns by suggesting that the visionary goal of beating the human world champions enables one to shape a research agenda formed by "a series of well-directed subgoals". Early RoboCup activities

[4] https://www.theguardian.com/theguardian/2011/may/12/deep-blue-beats-kasparov-1997.

illustrate this generative role. When RoboCup started, robotic teams were able to play rudimentary soccer games only. Measured with the yardstick of FIFA world championships, the performances of these robotic players were utterly unsatisfying. Nevertheless, their playing levels made initial benchmarks available to evaluate the next generation of teams participating in RoboCup leagues. Similarly, the improved skills of later generations of robotic teams made new benchmarks available for further research in the various RoboCup leagues. These playing benchmarks were supplemented with more traditional research targets identified in scientific symposia, concerning task-oriented visual feature selection, behavioral pattern modelling, coordinated action planning, and other skills making soccer playing possible.

There is no end in sight for improving robotic soccer capabilities using these various benchmarks from the vantage point of good human soccer playing levels. Hence, the RoboCup long-term goal is sensibly construed as an organizing or regulative idea for inquiry. This regulative idea is productive of a series of meaningful scientific and technological subgoals to pursue, independently of whether a robotic soccer team will eventually beat the soccer world champions.

RoboCup pioneers suggested a simile between the productive role of their visionary sports objective and the structure of the space program that President John F. Kennedy announced in a 1961 address to the US Congress by reference to the ultimate goal of "landing a man on the Moon and returning him safely to the Earth". The accomplishment of relatively more modest subgoals—from the development of suitable fuel and missile technologies to medical interventions supporting human life in microgravity conditions—enabled one to fulfil prerequisites for manned missions to the Moon. The achievement of these subgoals made a variety of new scientific knowledge, technologies and systems readily available to industry and other academic research areas. Accordingly, the ambitious long-term goal of the US space program in the 1960s fertilized many areas of science, technology and industrial R&D, before and independently of the walk that astronauts Armstrong and Aldrin took on the lunar surface on July 20, 1969.

Three different processes identifiable in the narratives concerning both RoboCup and the Apollo program are *Framework Building*, *Subgoaling*, and *Outcome Appraisal*. Framework Building is the process of identifying long-term goals that are well beyond state-of-art in both science and technology, along with scientific and technological tools that are initially suggested as appropriate means to pursue those goals. Subgoaling is the process of identifying and pursuing a series of increasingly ambitious objectives that are coherent with the general framework. Unlike the long-term goals of the research program, subgoals are expected to be achievable, measurable and rewarding in the short term. Outcome appraisal is the process of identifying and periodically applying suitable evaluation criteria that enable one to appreciate the contributions, if any, that the research program is making to science, technology and industry by pursuing its feasible subgoals.

The two processes of framework building and subgoaling that are involved in machine intelligence research programs on games appear to straightforward variations on two-process models of *scientific* inquiry (Godfrey-Smith 2003, pp. 117–121). Influential models in this class are Thomas Kuhn's model of normal science functioning, whose paradigm provides a general framework for scientists' daily puzzle solving activities, Imre Lakatos' model of scientific research programs, guiding the search for a series of models characterized by increasing predictive and explanatory powers, and Larry Laudan's model of research traditions, contributing to orient scientists in their selection of scientific allegiances and their choice of problems to solve.

The outcome appraisal process of a machine intelligence research program based on games, with its periodic evaluations of obtained results, affords a more interesting case study for the methodology of scientific and technological inquiry. The markedly multidisciplinary character of RoboCup and other research programs on machine intelligence games, their combination of scientific, technological, and industrial objectives, in addition to their seeming lack of useful technological applications, suggests that an intricate system of rewards and penalties must be taken into account in any sensible attempt to identify and appraise their outcomes.

So far, more than 300 papers of approximately 800 authors were published, regarding research in various AI areas developed for the RoboCup competition (https://www.robocup.org/research). Furthermore, RoboCup concepts and best practices inspired and contributed to promote similar events and initiatives, such as the EU project RoCKIn. This project aimed to encourage advancements in robotics and AI by designing new competitions and by introducing shared benchmarks enabling one to measure and rank the involved technologies and systems (http://rockinrobotchallenge.eu/). In (Lima et al. 2017) it is emphasized that RoCKIn competition design efforts extended the research topics involved in RoboCup, while retaining its approach in terms of broad regulative goals and defining outcome appraisal criteria for new competition challenges.

11.3 Outcome Appraisal, Double Allegiance and Regional Rewards

The development of robotic soccer teams provides an integrated task for a variety of machine intelligence research areas, including real-time sensor fusion, reactive behavior, strategy acquisition, learning, real-time planning, multiagent systems, context recognition, vision, strategic decision-making, motor control, intelligent robot control, and many more.[5] Identifying what RoboCup is contributing to achieve the specific objectives of each and every one of these research areas (in other words, their regional objectives, which may not be shared across the involved research areas) is a sensible criterion to use in order to appraise the overall outcomes of this research project. Qualitative estimates of RoboCup contributions to regional objectives of AI and robotics research must be identified by selective reference to, e.g., techniques for team coordination and distributed planning, temporally

[5] http://www.robocup.org/objective/.

constrained sensor fusion and visual processing, motion detection and modeling, efficient solutions of optimization problems with uncertain and incomplete data (Visser and Burkhard 2007). In their analysis of various regional contributions, Visser and Burckhard point out that scientists participating in RoboCup usually combine RoboCup research with other machine intelligence inquiries, and that they regard the exchange of ideas during tournaments as a valuable source of insights for their future work. This layered combination of scientific aims and activities reflects the *double allegiance* of most participating scientists. Their professional identities, career expectations and research ambitions are not centered around RoboCup. Rather, these features of their profile are defined by selective reference to those traditional AI and robotics research areas where their primary allegiance lies. Their continued commitment to RoboCup and other research programs based on machine intelligence games typically depends on the fruitfulness of their secondary allegiance activities for their primary allegiance goals. Accordingly, heterogeneous criteria converge in the appraisal of RoboCup subgoaling activities, which are inherited from the respective primary allegiances of contributing scientists. The imported appraisal criteria influence individual scientists in their periodic decision to renew or else to waive their secondary allegiance commitments.

The history of computer chess research illustrates how appraisal criteria reflecting the primary allegiance of participating groups of scientists affects the evolution of research programs on machine intelligence games. Surprisingly enough, at a time when the performances of computer chess programs were dramatically improving, some groups of scientists withdrew their participation from computer chess research. They were motivated to do so by diminishing returns for their primary allegiance. Let's see.

The development of the Deep Thought chess program was a milestone in the history of computer chess. In a panel on computer chess held at the 1991 International Joint Conference on Artificial Intelligence, Feng-hsiung Hsu, one of Deep Thought designers, pointed out that inputs from human chess experts should not be invariably trusted for the purpose of improving computer chess performances. And he went on to claim that various changes to Deep Thought heuristic procedures suggested by human experts had been found to affect negatively its performances. Therefore, he concluded, human expert inputs should be limited; and knowledge for computer chess playing is more effectively and efficiently generated on a statistical basis rather than relying on symbolic learning or explicit domain modeling (Levinson et al. 1991, p. 548). This methodological change paid off in the light of the ensuing victory of the Deep Blue chess program over Kasparov in 1997. These achievements, however, were relatively unimportant from the standpoint of Allen Newell and Herbert Simon's primary allegiance to the design of computer programs to simulate human cognitive processes. These scientists prized computer simulations of human chess playing mostly as sources of psychological explanations of underlying mental processes. From this perspective, victories of computer chess programs over human chess masters which provided no insight into human cognitive processes were uninteresting (Cordeschi 2002, p. 182).

John McCarthy was similarly dissatisfied, albeit for different motives, with the growing emphasis on tournament success and the use of more powerful hardware resources in computer chess. McCarthy pursued computer chess inquiries insofar as chess heuristic strategies were a key to understanding how any cognitively or computationally limited system—of either biological or artificial *genera*—might successfully cope with complex information processing tasks (Cordeschi 2002, pp. 200–201). Clearly, he did not prize chess tournament victory in and for itself, but only as a means to make progress towards his primary allegiance goals.

McCarthy expressed his reservations about developments in computer chess inquiries in his review of a book devoted to the history of chess programs. He claimed there that tournament success had come with a relevant scientific cost. Drawing on the simile attributed to the mathematician Alexander Kronrod—according to which "Chess is the *Drosophila* of artificial intelligence"—McCarthy complained that "computer chess has developed much as genetics might have if the geneticists had concentrated their efforts starting in 1910 on breeding racing *Drosophila*. We would have *some* science, but mainly we would have very fast fruit flies" (McCarthy 1997, p. 1518). He went on to argue that computer chess inquiries were no longer a good source of insights into general principles and mechanisms that enable cognitively bounded agents to cope with computationally complex tasks. In particular, he identified a chief motive of diminishing returns for his primary allegiance goals in the use of increasingly powerful hardware resources and the corresponding possibility of chess programs to explore more and more moves ahead in the game. Chess programs exploiting this feature had better chances to win than earlier chess programs relying on less powerful hardware and more limited search ahead strategies. However, the heuristic strategies of these better players failed to shed light on mechanisms that more limited computational agents use to cope with chess and other computationally complex problem-solving tasks.

11.4 Computer Game Research from Chess to Go and beyond

As chess was becoming less interesting for the primary allegiance of many researchers, the game of Go started to be seen by many as a new *Drosophila* of AI. Several motivations led AI researchers to focus on this game as their secondary allegiance. First, the complexity of the game was a key factor, as the search-space of Go is several orders of magnitude higher than chess (about 10^{170} legal moves, as calculated by John Tromp).[6] Another key reason had to do with the visual nature of the game. A single stone has no special value, for what matters most are groups of stones and the areas they enclose. Thus, a chief problem in Go is the recognition of visual patterns, which is another traditional and still leading research topic in

[6] https://tromp.github.io/go/legal.html.

AI. Accordingly, algorithmic approaches to Go might give insights about both human and artificial pattern recognition and perceptual capabilities (Zobrist 1969).

Superhuman performances at Go were eventually achieved by *AlphaGo*, developed by *DeepMind*, an AI system which defeated Lee Sedol, one of Go top-ranked players. The training method for *AlphaGo* relied on a dataset of moves from professional player matches, which were then annotated by Go experts (Silver et al. 2016). Soon after, *DeepMind* developed a novel Go agent, *AlphaGo Zero*, whose training relied solely on *reinforcement learning*, without using human expert inputs. *AlphaGo Zero* won over *AlphaGo* and reached a higher level in the *ELO* rating system[7] (Silver et al. 2017). The absence of humanly generated training data enabled this agent to produce novel strategies, which were previously unseen in Go tournaments.

DeepMind later developed another agent, *AlphaZero*, capable to play multiple games: chess, Go and Shogi. Without changing architecture, this system defeated top ranked artificial agents using no humanly generated training data (Silver et al. 2018). *AlphaZero* took advantage of techniques successfully employed with *AlphaGo Zero*. In the light of *AlphaZero* achievements, the techniques originally developed within a Go research program are a significant novel result flowing from its subgoaling process. Indeed, these techniques enabled one to develop more malleable agents. And the latter might be profitably used to deal with a variety of real-world tasks requiring flexibility and adaptation.

Poker is another game providing a fruitful framework to advance research on traditional AI topics. Poker and Go search spaces are comparable in size. Unlike Go, however, Poker is an imperfect information game: players cannot directly appraise its states, insofar as some relevant information is hidden from each one of them (Billings et al. 1998). Several traditional AI research problems are strictly related to Poker: the modelling of *multiple competitive agents*, as each player plays against all other players; *risk management*, which involves choosing strategies for betting and predicting consequences; *agent modelling*, as each player has to model the behavior and strategies of opponents; development of *deception strategies* and detection of *unreliable information*, insofar as the game allows one to deceive opponents and requires one to uncover their bluffs. Thus, the game of Poker appears to be a suitable domain for secondary allegiances of AI scientists, as it enables one to develop technologies and systems for real world problems involving multiple competitive agents, risk management, agent modelling, and so on.

Relevant Poker research has been concerned with a variant called *Texas Hold'em Poker*. In this variant, players have no betting limit, and only two players challenge each other at each hand. Significantly, an artificial agent for *No-Limits Heads up Texas Hold'em Poker* called *Libratus* challenged individually and won against five top professional players (Brown and Sandholm 2017). After victories achieved in 1vs1 matches, new research moved on to consider games of full *No-Limit Texas*

[7] The *ELO* rating system measures the skill level of each player in a game. Initially introduced for chess by Arpad Elo, a physic professor at Marquette University and a chess master player, it was then applied to other games such Go, Poker, Scrabble and various video-games.

Hold'em Poker, involving six players at each hand. The Poker agent *Pluribus*, jointly developed by *Facebook* and *Carnegie Mellon University*, won both in a table with five Poker top-players and in matches with five *Pluribus* agents versus a single human player (Brown and Sandholm 2019). *Pluribus* employed techniques developed for *Libratus*, refined to deal with more opponents.

Results obtained in Poker AI pave the way to the development of real-world applications which require one to deal with hidden information and deception exploitation. Cyber-security and negotiation applications are significant cases in point. *Facebook* pointed out that techniques developed for *Pluribus* are not special to Poker, and that no prior expert knowledge is needed to exploit these techniques. In the light of these considerations, *Pluribus* was (somewhat more controversially) identified as a stepping stone towards a more general AI infrastructure for multi-agents real task environments, which is claimed to be the ultimate goal of the whole research effort.[8]

Research into AI games has additionally branched out towards AI systems capable to play video-games. The latter require one to model continuous environments, in contrast to the turn-based nature of classical games. For this reason, video-games provide a valuable framework to investigate methodologies for developing continuous environment real-world applications. The OpenAI project, whose long-term goal is to move toward an Artificial General Intelligence,[9] integrates the *Arcade Learning Environment* (*ALE*) in its framework OpenAI *Gym*. *ALE* provides over 500 *arcade* games developed for the *Atari 2600* console, which was commercially distributed in 1977. These games are far simpler than present-day video-games, but they already pose challenging problems for AI. *ALE* is used as a problem-solving environment enabling one to move towards the OpenAI long-term goal by achieving the significant subgoal of agents capable of engaging into a wide variety of video-games. These agents are trained on a fraction of *ALE* video-games, and their performance is evaluated on other, previously unseen games. Accordingly, agents must prove their ability to engage into games involving widely different rules and objectives (Bellemare et al. 2015). The achievement of this subgoal within the OpenAI research program is expected to lead—in the way of application-oriented and industrial payoffs—to agents showing forms of adaptation and flexibility that are needed in a variety of continuous environment real-world domains.

AI research based on *arcade* video-games reached an important result in 2013, when *DeepMind* developed a single agent capable of playing at the level of expert human players on seven *Atari* games (Mnih et al. 2013). In order to achieve this result, a novel *reinforcement learning* method, called *Deep-Q-Learning*, was developed. This method and its later variants are state-of-the-art methods for *reinforcement learning problems*, which are applied in different real-world domains, including resource management (Mao et al. 2016) and robotics (Levine et al. 2016).

---

[8] https://ai.facebook.com/blog/pluribus-first-ai-to-beat-pros-in-6-player-poker/.

[9] https://openai.com/about/.

Moreover, this method laid the basis for techniques subsequently adopted in *AlphaGo* and *AlphaZero*.

In addition to pursuing the long-term goal of building an *Artificial General Intelligence*, video-games research is developing methodologies for even more complex games. In particular, *Real-Time Strategy* (*RTS*) games have been used as sources of fruitful techniques and testbeds for investigating major AI problems. Players of *RTS* games have to manage military units and buildings, harvesting resources to produce them, in order to destroy the army of one or more opponents. *RTS* games are meaningfully related to many real-world problems for artificial agents: *limited time* to choose an action; agents acting *simultaneously*; *durative* actions; *partial observation* of the environment; *vast state spaces*, that are even more extended than Chess, Go or Poker (Ontañon et al. 2013). Recent research on *RTS* focused on the game *StarCraft II* (Blizzard 2010), which enables one to address significant challenges for *reinforcement learning* methodologies, and paved the way to the introduction of *SC2LE*, a learning framework based on this game (Vinyals et al. 2017). Among other *RTS* challenges, *StarCraft II* requires one to address the problem of balancing in-game *micro management* (control of each individual unit) and *macro management* (long-term decision-making on resource investment), which are strictly related to specific real-domain problems. In early 2019, *DeepMind* developed *AlphaStar*, an agent that achieved a significant victory against *MaNa*, one of world top players of *StarCraft II*. The *AlphaStar* agent employed a variety of known techniques for AI game-playing agents, adding a novel multi-agent learning method (Vinyals et al. 2019), which enables the agent to model long sequences of actions and to take long-term decisions. This architecture may prove useful for real-world tasks involving long-term planning, such as weather forecasting or language understanding.

11.5 System Integration and Industry, Trading and Sharing

As the involvement of firms like Google and Facebook in projects concerning Go, video-games and Poker suggests, rewards expected to flow from research programs on machine intelligence games are not limited to those that are prized from the perspective of the primary research allegiances of participating scientists. The importance of boosting industrial R&D and of feeding the pipeline between academic research and industry were already emphasized in the RoboCup manifesto as major criteria to appraise the success of research programs on games like robotic soccer playing that are of no practical importance in themselves. Technologies originally developed to enable cooperative behaviors of robotic soccer players are applicable to develop cooperating teams of service robots for which markets are steadily expected to grow,[10] including surveillance robots operating on museum grounds and

[10] http://www.ifr.org/service-robots/statistics/.

other public areas, logistics and warehousing robots, let alone platoons of self-driving cars. More generally, the engineer's heuristic thinking is crucially needed to map into useful real-world applications technologies and systems that are initially oriented to playing games.

Domain transfer problems arising from the absence of direct application in useful domains emerge in connection with other striking feats of AI and robotic research. There is more than meets the layperson's eye in the captivating performances of the humanoid robot ASIMO or the cheerfully marching bands of robotic wind instruments players. Indeed, the 130 patents that Honda obtained on account of research work for ASIMO enabled this company to develop useful real-world applications, including assisted mobility devices designed to help persons with reduced leg functionalities due to age, accidents or illnesses.[11] Another example of promising domain transfer is IBM's Watson system, which was designed to achieve superhuman skill in a popular US television quiz called "Jeopardy!". This quiz is suitable to address issues in major AI research areas, such as natural language processing and knowledge representation. Watson eventually won the 1,000,000$ grand prize by defeating two game champions. IBM subsequently shifted the application domain to the ethically most sensitive area of medical diagnosis and treatment, developing on the basis of Watson a new system called *Watson Health*.[12]

To distinguish between groups of scientists participating in machine intelligence research programs based on games, we used so far only lists of topics compiled by AAAI or by technical committees of RAS (Robotics and Automation Society[13]). There are many additional properties of research groups one may informatively use for this clustering purpose. Research groups differ from each other in the way of their preferred technological platforms, preferred application domains and research methodologies, the supply industries used for the procurement of specific components and subsystems, and so on. The research communities that one may identify along these various dimensions are comparable to a mosaic of subcultures. Interactions within each subculture are typically more extensive than interactions between subcultures. However, mutually fruitful exchanges and transactions arise from interactions between subcultures. Clearly, research programs based on AI games, in addition to robotic team games, offer significant environments for these exchanges and transactions.

To identify mechanisms governing scientific and technological transactions within a markedly varied landscape of research interests, Peter Galison suggested that one may profitably learn "from the anthropologists who regularly study unlike cultures that do interact, most notably by trade. Two groups can agree on rules of exchange even if they ascribe utterly different significance to the objects being exchanged; they may even disagree on the meaning of the exchange process itself.

---

[11] http://www.gizmag.com/honda-stride-management-assist/23512/.

[12] https://www.newscientist.com/article/mg21528796-400-watson-turns-medic-supercomputer-to-diagnose-disease/.

[13] http://www.ieee-ras.org/technical-committees.

Nonetheless, the trading partners can hammer out a *local* coordination despite vast *global* differences" (Galison 1997, p. 783). Galison himself explored these transactions in the history of physics, by reference to trade practices between the partly autonomous practices of physical instrumentation, experimentation, and theory.

In research programs based on machine intelligence games, exchanges between subcultures are not limited to trading of knowledge, technologies and systems. Standardization plays a methodologically significant role there, insofar as it enables one to alleviate methodological problems affecting solitary software development and robot construction practices. Typically, the behavior of robots constructed in one research lab depends on a wide variety of individual design choices and implementation details which concern, e.g., robot size, sensor systems, selection of actuators and balanced composition of their hardware and software components. Therefore, one is hardly pressed to find informative benchmarks to compare in depth the task performances of robots constructed in one's own lab with same task performances of robots constructed elsewhere on the basis of different design and implementation choices. RoboCup sharing of standard industrial platforms in both Simulation and real robotic soccer leagues provides some means to alleviate this methodological problem, just like the practice of making programs publicly available in the RoboCup Simulation league.

In conclusion, a complex system of rewards and penalties is required to appraise the outcomes of a research program based on machine intelligence games. This system involves regional interests in academic research, double allegiance of participating scientists, scientific and technological advances in machine intelligence, and their translational promises in the way of industrial applications in domains of practical interest. In addition to these items, one should consider a wide range of ancillary benefits leveraging on the allure of machine intelligence games. These benefits include the pedagogic payoff of attracting wider cohorts of science and engineering students, motivating effects for junior scientists, and the outreach of dissemination actions using machine intelligence games, which are expedient to engage audiences that are not familiar with AI and robotics. An understanding of this composite and intricate system of rewards and penalties is also needed on the part of individual scientists who decide whether to confirm or waive their secondary allegiance to some multidisciplinary research program based on AI games.

11.6 Concluding Remarks

Gamification is a recently coined word referring to the application of game design principles in non-gaming contexts (Robson et al. 2015). Gamification appears to be an increasingly popular strategy in the definition and pursuit of research programs in machine intelligence. For example, RoboCup's initial focus on soccer playing later expanded towards socially useful robotic tasks and related competitions between robotic systems. The long-term goal of RoboCup@home (www.robocupathome.org) is to develop autonomous robots which share with humans the physical space of their homes and carry out there a variety of household and assistive tasks. RoboCup Rescue (www.robocuprescue.org) is a competition recruiting and comparing the search and rescue capabilities of robotic teams working in earthquake-stricken and other disaster areas.

Gamification of research activities is a strategy pursued by major research funding schemes too, notably including the EU Horizon 2020 program. The latter has called for the "organisation of robotic competitions to speed up the advance towards smarter robot, demonstrating progress in the field and raising the awareness of the general public towards intelligent robots[14]". Major military research funding agencies also pursue gamification of research in machine intelligence. Notably, from 2013 to 2015 the Defense Advanced Research Projects Agency (DARPA) of the US Department of Defense (DoD) promoted the Robotic Challenge, which was driven by the long-term goal of building autonomous robotic systems replacing human operators in dangerous task environments. The Robotic Challenge developed through robotic tournaments for task-level autonomy and operational capabilities in hazardous trial arenas. Even though these arenas were not directly identifiable with warfare scenarios, novel technologies and systems demonstrated there were endowed with many features of interest to armed forces and military industries. Accordingly, the system of rewards for appraising research programs based on games and competitions must be enriched to include military application payoffs. Similarly, the deliberation processes that scientists engage into to decide whether to confirm or to waive their allegiance to research programs based on machine intelligence games must take into account the potential dual use of their research products and related ethical issues (Tamburrini 2016).

Visionary goals for machine intelligence games have not been invariably productive of fruitful scientific and technological research programs. A case in point is Turing's imitation game, also known as Turing Test, and the related ambitious objective of building a computer doing well in the game. The imitation game involves a human being, a computing machine, and another human being in the role of interrogator who communicates with the other two players by texting only. By asking them questions, the interrogator must discover which one of her interlocutors is the machine. A machine playing well the imitation game will bring its interrogators to make the wrong identification "often enough" after "sufficiently extended" periods of conversational exchanges (Turing 1950). The natural language processing capabilities that are required to play well the imitation game are a potentially rich problem-solving domain for machine intelligence inquiry. However, accomplishment criteria for the imitation game—unlike those for chess, soccer, Poker, Go and beyond—are rather vague and open to ample subjective disagreement. Moreover, the long-term goal of passing Turing's test—whatever this may precisely mean—has not given rise yet to a series of feasible subgoals enabling one to advance computer playing skills through competition and ranking. The annual competition named Loebner Prize was inspired to Turing's test, but its framework is clearly

[14] https://ec.europa.eu/programmes/horizon2020/en/draft-work-programmes-2016-17.

inadequate to provide significant benchmarks to measure, compare and rank the linguistic capabilities of participating systems (Shieber 1994; Powers 1998).

It is doubtful that Turing meant the imitation game as a regulative goal for a genuine research program on machine intelligence games. Arguably, Turing's intentions were more pedagogic in character: the imitation game enabled him to convey vividly to the general public the idea that computing machines would gradually become able to compete and cooperate with humans in a wide variety of symbol-processing activities (Gandy 1996, p. 125; Cordeschi and Tamburrini 2015). And in later articles, written shortly before his tragic death, Turing actively suggested to use chess and other social games as generators of research programs in machine intelligence (Turing 1953). In particular, he was careful to distinguish between different levels of chess playing skills, thereby suggesting a series of subgoals for computer chess inquiries and the ranking of chess programs (Turing 1953; Turing 1992, pp. 163–164).

As AI systems became capable to achieve superhuman skills in some given game, researchers often shifted their attention towards more complex games, say, from chess to Go, and then from Go to Poker. Moreover, we have pointed out that techniques developed for simpler games often aid researchers to deal with more complex games. In similar cases, resolving simpler games is a subgoal, whose achievement enables one to address subsequent and more challenging subgoals. Methods developed to deal with some particular game often turned out to be less domain-independent than one may have expected at first. The *deep-Q-learning* technique is emblematic in this sense, as it was applied to a wide variety of different domains, ranging from robotics to chemistry.

Machine intelligence games (notably including Chess, Go, Poker, robotic games and video-games) have been used here as a collective *Drosophila* for the methodology of AI and robotics. Indeed, research programs that are based on machine game playing offer a vantage point on central features of the complex systems of rewards, penalties and appraisal criteria that come into play in machine intelligence inquiry. In particular, these appraisal criteria must make room for what has been called the double allegiance of scientists, and the benefits for their primary allegiance flowing from work done on a multidisciplinary research program that scientists espouse as their secondary allegiance. Similarly, the sharing, trading and standardization benefits flowing from multidisciplinary research programs should not be belittled in research areas like AI and robotics, where intersubjective criteria to identify what counts as a result are not easy to come. These various ingredients should be properly taken into account to develop more informative models of systems of rewards for scientists (Strevens 2003; Weinsberg and Muldoon 2009; McKenzie et al. 2015; Baumberg 2018), which extend the classical methodological pictures of rewards and penalties for scientific and technological inquiry developed during the twentieth century.

References

Baumberg, J. (2018). *The secret life of science: How it really works and why it matters*. Princeton: Princeton University Press.

Bellemare, M. G., Naddaf, Y., Veness, J., & Bowling, M. (2015). The arcade learning environment: An evaluation platform for general agents. In *Proceedings of the 24th international conference on artificial intelligence*. Buenos Aires: AAAI Press.

Billings, D., Papp, D., Schaeffer, J., & Szafron, D. (1998). Poker as a testbed for AI research. In R. E. Mercer & E. Neufeld (Eds.), *Advances in artificial intelligence* (pp. 228–238). Berlin, Heidelberg: Springer.

Brown, N., & Sandholm, T. (2017). Libratus: The superhuman AI for no-limit poker. In *Proceedings of the twenty-sixth international joint conference on artificial intelligence, IJCAI-17*. Melbourne: IJCAI.

Brown, N., & Sandholm, T. (2019). Superhuman AI for multiplayer poker. *Science, 365*(6456), 885–890.

Cordeschi, R. (2002). *The discovery of the artificial. Behavior, mind and machines before and beyond cybernetics*. Dordrecht: Kluwer.

Cordeschi, R., & Tamburrini, G. (2015). Alan Turing e il programma di ricerca dell'Intelligenza Artificiale. In H. Hosni (Ed.), *Menti e macchine. Alan Mathison Turing a cento anni dalla nascita* (pp. 85–124). Pisa: Edizioni della Normale.

Galison, P. (1997). *Image and logic: A material culture of microphysics*. Chicago: University of Chicago Press.

Gandy, R. O. (1996). Human versus mechanical intelligence. In P. Millican & A. Clark (Eds.), *Machines and thought* (pp. 125–136). Oxford: Clarendon Press.

Godfrey-Smith, P. (2003). *Theory and reality: An introduction to the philosophy of science*. Chicago: University of Chicago Press.

Kuhn, T. S. (1962). *The structure of scientific revolutions*. Chicago: University of Chicago Press.

Lakatos, I. (1970). Falsification and the methodology of scientific research programmes. In I. Lakatos & A. Musgrave (Eds.), *Criticism and the growth of knowledge* (pp. 91–161). Cambridge: Cambridge University Press.

Laudan, L. (1977). *Progress and its problems: Toward a theory of scientific growth*. Berkeley: University of California Press.

Levine, S., Finn, C., Darrell, T., & Abbeel, P. (2016). End-to-end training of deep visuomotor policies. *Journal of Machine Learning Research, 17*(1), 1334–1373.

Levinson, R., Feng-hsiung Hsu, Marsland, T. A., Schaeffer, J., Wilkins, D. E. (1991). Panel on computer chess. In J. Mylopoulos & R. Reiter (Eds.), *Proceedings of the 12th international joint conference on artificial intelligence*. Sydney, San Francisco: Morgan Kaufmann.

Lima, P., Nardi, D., Kraetzschmar, G., Bischoff, R., Matteucci, M. (2017). RoCKIn and the European robotics league: Building on RoboCup best practices to promote robot competitions in Europe. In S. Behnke (Ed.), *RoboCup 2016: Robot world cup XX* (pp. 181–192). Leipzig: Springer.

Mao, H., Alizadeh, M., Menache, I., & Kandula, S. (2016). Resource management with deep reinforcement learning. In *Proceedings of the 15th ACM workshop on hot topics in networks*. Atlanta: ACM.

Marr, D. (1977). Artificial intelligence – A personal view. *Artificial Intelligence, 9*(1), 37–48.

McCarthy, J. (1997). AI as sport. *Science, 276*(5318), 1518–1519.

McKenzie, A. J., Himmelreich, J., & Thomson, C. (2015). Epistemic landscapes, optimal search, and the division of cognitive labor. *Philosophy of Science, 82*(3), 424–453.

Mnih, V., Kavukcuoglu, K., Silver, D., Graves, A., Antonoglou, I., Wierstra, D., Riedmiller, M. (2013). *Playing Atari with deep reinforcement learning, "arXiv"*.

Ontañon, S., Synnaeve, G., Uriarte, A., Richoux, F., Churchill, D., Preuss, M. (2013). A survey of real-time strategy game AI research and competition in StarCraft. *IEEE Transactions on Computational Intelligence and AI in Games, 5*(4), 293–311.

Powers, D. M. W. (1998). The total turing test and the Loebner prize. In *Proceedings of the joint conferences on new methods in language processing and computational natural language learning*. Sydney: Association for Computational Linguistics.
Robson, K., Plangger, K., Kietzmann, J.H., McCarthy, I., Pitt, L. (2015). Is it all a game? Understanding the principles of gamification. *Business Horizons, 58*(4), 411–420.
Shannon, C. E. (1959). Programming a computer playing chess. *Philosophical Magazine, 41*(314), 256–275.
Shieber, S. M. (1994). Lessons from a restricted turing test. *Communications of the ACM, 37*(6), 70–78.
Silver, D., Schrittwieser, J., Simonyan, K., Antonoglou, I., Huang, A., Guez, A., et al. (2016). Mastering the game of go with deep neural networks and tree search. *Nature, 529*(7587), 484–489.
Silver, D., Hubert, T., Schrittwieser, J., Antonoglou, I., Lai, M., Guez, A., et al. (2017). Mastering the game of go without human knowledge. *Nature, 550*, 354–359.
Silver, D., et al. (2018). A general reinforcement learning algorithm that masters chess, shogi, and go through self-play. *Science, 362*(6419), 1140–1144.
Strevens, M. (2003). The role of the priority rule in science. *Journal of Philosophy, 100*(2), 55–79.
Tamburrini, G. (2015). Machine intelligence sports as research programs. *Paradigmi*, 163–177.
Tamburrini, G. (2016). On the ethical framing of research programs in robotics. *AI & Society, 31*(4), 463–471.
Turing, A. M. (1950). Computing machinery and intelligence. *Mind*, (59), 433–460.
Turing, A. M. (1953). Digital computers applied to games. In B. V. Bowden (Ed.), *Faster than thought. A symposium on digital computing machines* (pp. 161–185). London: Pitman & Sons.
Turing, A. M. (1992). *Mechanical intelligence (collected works of A. M. Turing)*. Amsterdam: North-Holland.
Vinyals, O., Ewalds, T., Bartunov, S., Georgiev, P., Vezhnevets, A.S., Yeo, M., et al. (2017). StarCraft II: A new challenge for reinforcement learning. *Machine Learning*. arXiv:1708.04782.
Vinyals, O., Babuschkin, I., Chung, J., Mathieu, M., Jaderberg, M., Czarnecki, W., et al. 2019. AlphaStar: Mastering the real-time strategy game StarCraft II. Available at: https://deepmind.com/blog/alphastar-mastering-real-time-strategy-game-starcraft-ii/.
Visser, U., & Burkhard, H. D. (2007). RoboCup: Ten years of achievements and future challenges. *AI Magazine, 28*(2), 115–132.
Weinsberg, M., & Muldoon, R. (2009). Epistemic landscapes and the division of cognitive labor. *Philosophy of Science, 76*(2), 225–252.
Zobrist, A. L. (1969). A model of visual organization for the game of go. In *Proceedings of the may 14–16, 1969, spring joint computer conference*. Boston: ACM.

Chapter 12
Robots and Bionic Systems as Experimental Platforms for the Study of Animal and Human Behaviour

Edoardo Datteri

Abstract Since the early decades of the XX century, robots have been often used as experimental platforms for the study of animal and human behaviour. This article offers an overview of these epistemic uses with a special focus on contemporary biorobotics and bionics. Some experimental procedures for the study of the behaviour of animals and humans, involving fully artificial and hybrid bionic systems, are reconstructed on the basis of examples drawn from the scientific literature. Some conditions under which they can produce genuine knowledge about living systems are identified and discussed.

Keywords Biorobotics · Biomimetics · Simulations · Robotic models · Synthetic method

12.1 Introduction

What can robotic models offer to the study of animal and human behaviour? The construction of automata has informed theorizing on the behaviour of living organisms for centuries (Riskin 2016; Truitt 2016). However, since the early decades of the XX century, robots have also played experimental roles, i.e., they have been involved as key components in experiments for the study of individual and social behaviour. In a large class of robot-supported studies belonging to so-called biorobotics, theoretical conclusions about the behaviour of living systems are drawn from observation and analysis of robotic or hybrid systems. This article offers an over-

E. Datteri (✉)
RobotiCSS Lab, Laboratory of Robotics for the Cognitive and Social Sciences,
Università degli Studi Milano-Bicocca, Milan, Italy
e-mail: edoardo.datteri@unimib.it

S. Chiodo, V. Schiaffonati (eds.), *Italian Philosophy of Technology*, Philosophy of Engineering and Technology 35,
https://doi.org/10.1007/978-3-030-54522-2_12

view of contemporary biorobotics, with a descriptive goal—to reconstruct some biorobotic experimental procedures for the study of the behaviour of living systems—and a methodological goal—to identify some conditions upon which the validity of these experimental procedures rests.

The role that automata have played for centuries in shaping mechanistic conceptions of life and inspiring theories of animal and human behaviour has been occasionally discussed in the philosophical literature. However, few studies have been devoted to the analysis of the *experimental* roles played by robots in the life sciences. Philosophers have mostly focused so far on the reconstruction and analysis of the role played by non-robotic machines—computer simulations *qua* machines implementing theoretical models of biological systems—in the study of biological behaviour (Weisberg 2013; Winsberg 2010). There are exceptions: methodological analyses of the experimental use of robots are offered in (Cordeschi 2002; Datteri 2017a, b; Datteri and Tamburrini 2007; Pfeifer and Bongard 2006; Tamburrini and Datteri 2005; Webb 2001). Interestingly, these studies focus on a particular experimental robot-supported method often called "synthetic method" which was advocated, among others, by pioneers of Artificial Intelligence Allen Newell and Herbert Simon (Simon and Newell 1962). The synthetic method, discussed below, is surely one of the key methods in biorobotics. However, especially in recent times, novel robot-supported experimental procedures for the study of animal and human behaviour have emerged. They will be illustrated and discussed in this article.

The structure of this article is as follows. In Sect. 12.2 it will be suggested that robots may be used to promote mechanistic conceptions of animal behaviour and to test specific behavioural hypotheses. In Sect. 12.3, biorobotic methodologies involving fully artificial, non-hybrid systems will be analysed. Section 12.4 will deal with variants of the synthetic method involving hybrid bionic systems connecting biological tissues with electromechanical devices.

12.2 From Inspirational Sources to Experimental Testbeds

Observation of machines influenced theorizing about the behaviour of living systems well before the XX century. As often pointed out, the development of mechanistic conceptions of life in the modern and Enlightenment era was informed by observation of hydraulic machines, clocks, automated theatrical representations, and humanoid automata. Machines, in these cases, inspired the general idea that life and behaviour could be produced by mechanisms. In other cases, machines inspired the formulation of more specific hypotheses about the structure of the mechanisms underlying particular aspects of living system behaviour. Riskin (2016) reports that Kenelm Digby, an English Cartesian thinker, proposed in the first half of the XVII century that the structure of the pumping apparatus of plants was internally similar to the mechanism that raised water from the Tagus River to supply the Alcazar palace in Toledo. Over and above supporting a mechanist conception of life, in this and other cases a hypothetical description of the structure of a mechanism governing the behaviour of a class of living systems (e.g., of the pumping mechanisms of plants) was derived from a description of the structure of a man-made mechanism (e.g., of a water supply mechanism). It is in this sense too that machines have often *inspired* the formulation of mechanistic hypotheses about life and behaviour.

The distinction between these two ways in which mechanical devices can inform theorizing about living system behaviour—supporting a mechanistic conception of life and behaviour, and inspiring the formulation of specific mechanistic hypotheses—can be found in more recent times too. As extensively discussed by Cordeschi (2002), the construction of, and subsequent observation and reflection on, automata was often taken in the first decades of the XX century as a basis to believe that many aspects of animal and human behaviour that were previously thought to be prerogative of living systems—including variability, purposeful behaviour, and learning—could nevertheless be explained, at least in principle, mechanistically. Grey Walter's tortoises famously contributed to reinforce the idea, later developed by Valentino Braitenberg (1986), that "simple" mechanisms—having few functional components and mutual connections—could produce variable, unpredictable, and seemingly purposeful behaviours (Grey Walter 1950). Thomas Ross' hydraulic machines and mechanical maze-searching rats, provided with elementary forms of memory, were taken as a basis to believe that learning phenomena could be reproduced in mechanisms (Cordeschi 2002). In addition to the mere support for a mechanistic stance towards life and behaviour, machines and robots have more recently shaped research on cognitive and neural mechanisms. For example, the fact that a carefully designed physical structure may enable robots with "simple" control mechanisms to perform complicated tasks has inspired the development of so-called embodied branches of cognitive science. Embodied and non-embodied approaches to cognitive science share a common mechanistic approach to the explanation of animal and human behaviour but differ in the structure of the mechanisms they postulate. Inspired by the behavioural skills of "simple" reactive robots with carefully crafted bodies (Brooks 1999), embodied approaches promote explanations which put greater emphasis on body mechanics than on control algorithms as determinants of animal and human behaviour (Pfeifer and Bongard 2006).

In the studies that will be discussed here, robots play a crucial role not (only) in the promotion of a general mechanistic stance on the generation of living system behaviour, but (also) in the formulation of specific mechanistic hypotheses. Moreover, the role played by robots is not only inspirational in the sense discussed above: in contemporary biorobotics, robots are deliberately built to be involved in experiments for the study of animal and human behaviour. Their behaviour is analysed in controlled experimental settings, and the results of the analysis are brought to bear on the plausibility of theoretical models formulated for specific explanatory or predictive purposes. Often, some features of the robots are selectively altered to address specific aspects of the research question or to test alternative explanations of the results. The experimental role of robots will be discussed in the following sections in connection with biorobotic models of animal behaviour and with bionic systems connecting biological tissues with electromechanical devices.

12.3 Biorobotics

12.3.1 Theoretical Models and Robotic Systems

In so-called biorobotics (Datteri 2017a; Webb 2008), theoretical conclusions about the behaviour of a living system are drawn from observation and analysis of a robotic system. What kind of theoretical conclusions may be obtained, what experimental procedures may be adopted, and under what conditions may these experimental procedures enable one to acquire genuine scientific knowledge about the target living system?

Observation of a robot in controlled experimental settings may lead to two kinds of scientific results about the target living system. It may inform one about the plausibility of a theoretical model of the target system (more on the notion of "theoretical model" below). Or, it may inform one about the behaviour that the target system would generate in some circumstances. The robot is useful, in this second case, especially when the behaviour of the target system cannot be directly observed for practical or theoretical reasons (e.g., because it is a future behaviour or because, as in the study discussed below, the target animal is extinct). Studies of the first kind are called model-oriented here (Datteri 2017a; Datteri and Schiaffonati 2019), and the experimental procedure they adopt has been often dubbed "synthetic method" (Cordeschi 2002). Studies of the second kind are called prediction-oriented here and, as clarified below, adopt a relatively different methodology. In order to understand the distinction between model-oriented and prediction-oriented studies, the structure of the experimental procedures adopted in the two cases, and the conditions of validity of the two methodologies, it will be useful to reflect on what relationship may hold between the robotic system involved in the experiment, the target biological system, and a theoretical model of the latter.

By theoretical model we mean here a set of assumptions about a system (Achinstein 1965). More specifically, theoretical model M characterizes target system B in terms of some properties and regularities holding among them. Suppose, for example, that B is a moth and M is a Braitenberg-like model of it (Braitenberg 1986). M models B in terms of *properties* characterizing, say, activation of left and right motor organs and the level of stimulation of left and right light sensors. It also charaterizes it in terms of *regularities* stating, say, that the level of left motor activation is directly proportional to the level of right sensor activation (and vice versa), and that left motor activation will make the system turn right (and vice versa). Affirming that M is a "good" model of B amounts to affirming that B's properties stand with one another in the relationships expressed by M—for example, that moths can be characterized in terms of left and right motor organs and level of stimulation of left and right light sensors, and in terms of regularities stating that left motor activation is directly proportional to the level of right sensor activation, and so on. Some theoretical models can be used to explain the behaviour of the target system—for example, to explain why moths seek light or why a particular moth eventually produced a particular trajectory. Some theoretical models may be used to predict the behaviour of the target system—for example, to predict that, in normal conditions, moths seek light or that a moth in particular circumstances will turn left or right, or produce a particular trajectory.

Implementing theoretical model M in a robotic system amounts to building a robotic system A which can be characterized in terms of properties standing with each other in the same relationships as those stated by M. For example, implementing the Braitenberg-like model sketched above in a robot amounts to building a system having sensors and motor devices, so that the level of activation of the left motor devices is proportional to the level of activation of the right sensor (and vice versa), and that left motor activation makes the system turn right (and vice versa). One thus obtains a robotic system A which implements theoretical model M of target system B. Note that one can accurately implement M in A regardless of whether or not M is a "good" model of B: one thing is to say that M is a good model of B, another thing is to say that A accurately implements M. Establishing the plausibility of M as a model of B and establishing whether M has been accurately implemented in A are completely different processes. Note also that A will surely display features which are not represented in M. Batteries, cooling fans, wi-fi devices for exchanging of data with external laptops, may be required for the Braitenberg-like system described above to work or to be used in the experiments, even though they are not represented in M.

This analysis gives rise to many questions concerning modelling and implementation which are not addressed here. No precise definition is given of the notions or "property" and "regularity". There is ample philosophical literature about the notion of theoretical model, about what makes a model explanatory, about the relationship between models and theories, between models and concrete systems (Magnani and Bertolotti 2017), and between models and mechanisms (Glennan and Illari 2018). Addressing these questions may be important to reflect about specific issues concerning biorobotics. However, the sketchy characterization of the concept of theoretical model and of the relationship between models, target system and robotic system provided here will suffice to illustrate the most relevant methodological differences between model-oriented and prediction-oriented studies, and to outline the conditions under which these two methods provide genuine knowledge about the target living system.

12.3.2 Model-Oriented and Prediction-Oriented Studies

In model-oriented studies (i.e., in studies adopting the model-oriented strategy), the behaviour of robotic system A is compared with the behaviour of the target system B in particular experimental settings. If the two behaviours match, one may conclude—on some auxiliary assumptions that are discussed below—that M is a plausible model of B. A mismatch may lead one to reject M as a model of B. As pointed out before, this strategy has been often called synthetic method (Cordeschi 2002). Note that A's reproduction of B's behaviour, *per se*, provides conclusive reasons to conclude neither that M is the only possible model of B nor that it has explanatory or predictive value.

In prediction-oriented studies (i.e., in studies adopting the prediction-oriented strategy), A's behaviour is interpreted as the behaviour that B would produce in the same conditions. This interpretive step is not supported by comparisons between A's and B's behaviours, as in the model-oriented strategy: indeed, as pointed out before, prediction-oriented studies are often carried out when independent information about B's behaviour is impossible to obtain. Vice versa, the interpretive step characterizing prediction-oriented studies is not taken in the model-oriented strategy: in a model-oriented study, one has no *a priori* reason to believe that A's behaviour will match B's behaviour. To be sure, what characterizes the model-oriented strategy is not the mere fact that A's and B's behaviours are compared with one another, but the fact that the outcome of the comparison is used to evaluate the plausibility of M. Indeed, the mere absence of a comparison is not a good reason to conclude that the analysis in question is prediction-oriented: it may be simply the case that data on B's behaviour are yet to be acquired. Moreover, machine outputs can be compared with B's behaviours for purposes other than theoretical model corroboration or rejection—for example, as discussed below, to test the predictive adequacy of a prediction-oriented system, with no claim whatsoever on the plausibility of the underlying theoretical model.

An example of a model-oriented study can be found in Grasso et al. (2000). The goal of the study was to explain how lobsters can reach sources of chemical plumes in the water, possibly released by organisms such as molluscs. The authors started with a very simple mechanistic hypothesis. It is reasonable to guess that chemical concentration will be high near the source and low at the periphery, so that local differences in chemical concentration will reflect the relative position of the source: if this is the case, the animal has just to climb the concentration gradient to reach the destination. Lobsters' antennae are chemical sensors. The authors therefore hypothesized that lobsters, to perform that task, steer towards the side corresponding to the higher concentration based on the following Braitenberg-like relationship: the higher is the concentration perceived at the left antenna, the higher is the velocity of the right motor organs of the animal, and vice versa.

This mechanistic hypothesis is intuitively plausible—at least, if one assumes that the chemical stream spreads in a regular way from the source. But water is turbulent, and turbulence may perturb the chemical stream at the periphery. If this is the case, local differences in chemical concentration in the periphery may be poorly informative of the position of the source, and the proposed mechanism will disorient the animal. Is the chemical stream regular enough to allow the simple mechanism proposed by the authors to guide the animal towards the destination?

To address this question, the authors built *RoboLobster*, a small underwater robot moving on wheels and provided with two chemical sensors. The control mechanism incorporates the hypothesis formulated above: each lateral chemoreceptor of the robot is connected to the motor, speeding it up when higher concentration is detected, as prescribed by the hypothesized mechanism. The idea is to put the robot in the water, immersed in a chemical stream propagating from a distant source: if the robot consistently succeeds in reaching the chemical source (thus matching lobsters' chemiotaxis abilities), then one may be induced to corroborate the mechanistic hypothesis under scrutiny. Otherwise, one may be induced to reject it. The authors carry out these experiments and obtain non-homogeneous results: the robot almost always reaches the source when starting at a distance of 50 cm from it, and almost always fails when starting at a distance of 1 m. Based on these results, the authors conclude that the chemical stream is not regular at the periphery of the stream, probably due to water turbulence. Now, lobsters outperform the robot in this task. For this reason, the authors conclude that the proposed mechanism is too simplistic and should be rejected as it stands: additional mechanisms must be identified to explain how lobsters find the right direction even when the plume is irregular.

The lobster study is an example of the model-oriented strategy: a theoretical model formulated to explain the behaviour of a living system is evaluated (rejected, in this particular case) based on a comparison between the behaviour of the target system and of a robot implementing the model. The goal of the study was chiefly to test a theoretical model of lobster behaviour. The *prediction-oriented* strategy, not conforming to the synthetic method, is exemplified by one of the studies reported by Long (2012). In 1821 a fossil of a giant vertebrate, later called *Plesiosaurus*, was discovered along the coast of the English Channel. The *Plesiosaurus* had four large flippers which were apparently shaped to do the hydrodynamic work of underwater wings, suggesting that they were all used to swim. Contemporary living tetrapods using appendages to swim (including penguins and sea turtles) use either pectoral or pelvic limbs, but not both, for propelling themselves. "Given the species richness of the Mesozoic radiation of tetrapods with apparent four-flippered propulsion, why do more recently derived aquatic tetropods use only two limbs for propulsion? Which permits better swimming performance, four flippers or two?" (Long et al. 2006, p. 21).

To address this question, John Long and his team built a robot, called *Madeleine*, able to swim with four or two flippers. The robot, using four and two flippers in different experimental sessions, was left free to swim in a pool while its speed, acceleration, and energy consumption were recorded. The results indicate that four flippers guarantee approximately the same top speed, that four flippers guarantee higher acceleration, and that two flippers guarantee lower energy consumption. From these results specifically concerning *Madeleine*, the authors drew some conclusions about the behaviour of *Plesiosauri* including "that the plesiosaurs, bearing four putatively-propulsive flippers ... were likely to have been ambush predators, taking advantage of the better acceleration of four flippers over two" (Long et al. 2006, p. 27). The reason why many aquatic tetrapods evolved into two-legged propulsion was identified by the authors in the lower energy consumption guaranteed in this condition.

To sum up, in this work, observation and analysis of robotic behaviour informed the authors about the behaviour (speed, energy consumption, acceleration) that the *Plesiosaurus* would have produced in the same conditions. Unlike the "synthetic method", the theoretical conclusions reached there are not drawn from comparisons between biological and robotic behaviour: the behaviour of the robot was interpreted as the behaviour that the target system would have generated in the same conditions.

12.3.3 Conditions of Validity

Under what conditions can one safely draw theoretical conclusions about model plausibility and target system behaviour from the results of a model-oriented and a prediction-oriented study, respectively?

Consider model-oriented studies first. Recall that, in the synthetic method, one implements theoretical model M in robot A and analyses A's ability to reproduce the behaviour of the target system B: matches or mismatches between the two are brought to bear on the plausibility of M as a model of B. However, A may succeed or fail to reproduce B's behaviour independently of whether M is a plausible model of B. The robot may fail to reproduce the behaviour of the target system even though the implemented model is a "good" model of it, simply because the robot implements it inaccurately (recall the remarks about implementation accuracy made in Sect. 12.3.1). Or, it may succeed to reproduce the behaviour of the target system even though M is a "bad" model of it, simply because the robot implements some *ad hoc* correctives which were not prohibited by M. These considerations could be clarified by a more precise account of what makes a robotic implementation accurate. Nevertheless, they suffice to illustrate a general methodological problem affecting the model-oriented procedure in biorobotics. Ideally, to safely draw conclusions about M's plausibility based on behavioural comparisons between A and B, one must assume that M is accurately implemented in A and that A implements no *ad hoc* corrective. If this is not the case, one must carefully reflect about whether the implementation inaccuracies and correctives found in A may significantly affect the behaviour of the robot.

Note that establishing implementation accuracy may be rather difficult in some cases, for reasons connected to the features of the theoretical model itself. A yes/no judgement on whether M is accurately implemented in A requires one to reflect on the relationship between the assumptions made in M and the characteristics of A. But the outcome of this evaluation process may be controversial if these assumptions are formulated in sketchy, vague or qualitative way. In some cases, it may even be difficult or impossible to decide whether particular characteristics of A conform to M, violate M, or are totally unconstrained by M. Biorobotic studies often start from theoretical models with qualitative, verbally expressed models such as "left motor speed is proportional to right chemoreceptor activation, and vice versa". Implementation, as pointed out, typically requires one to add details which were not mentioned in M. Some implementation details (the addition of a wi-fi module to the robot, for example) can be reasonably regarded as totally unrelated to B, and totally unconstrained by any plausible version of M. Other implementation details, such as the addition of sensory processing modules for chemical gradient detection, though literally not mentioned in M, may be regarded as implicitly constrained by it. Should one conclude that A inaccurately implements M, as literally stated? Or, that A accurately implements M because it implements the implicitly constrained features too? These considerations, which would deserve a separate analysis, suggest that evaluating implementation accuracy—a key methodological requirement of model-oriented studies—may be hard or impossible in some cases, for reasons connected to the way M is formulated.

In prediction-oriented studies, A's behaviour is interpreted as the behaviour that B would produce in the same conditions. One may reasonably believe that M must be a predictively valid model, and that M must be accurately implemented in A, to make such a predictive use of the latter. For, if M was predictively invalid, there would be no reason to believe that an accurate implementation of it would predict the behaviour of the target system. And, if M was inaccurately implemented in A, though being predictively valid, there would be no reason to interpret A's behaviour as predictive of B's behaviour. It may be concluded from these remarks that both conditions—M's predictive validity and implementation accuracy—must be separately met to make a predictive use of A. However, as discussed in the philosophy of simulation literature in connection with predictive uses of computer simulations (Winsberg 2010), there is no reason to believe that both conditions must hold, and consequently that sanctioning predictive adequacy of A requires one to separately evaluate model predictive validity and implementation accuracy. For, in principle, A may be predictively adequate even though M is predictively invalid and inaccurately implemented in A simply because the effects of the two conditions cancel each other: one may establish A's predictive adequacy, therefore, simply based on a history of predictive successes.

12.4 Bionics

12.4.1 A Taxonomy of Experimental Methods

Bionic technologies connect electronic or robotic devices with living tissues via invasive (e.g., microelectrode arrays) or non-invasive (e.g., electroencephalographic electrodes) interfaces. They may play an important role in restoring sensory and motor functionalities of impaired people via prosthetic arms controlled by the central or peripheral nervous system. In addition, it has been claimed that "the general strategy … of using brain-derived signals to control external devices may provide a unique new tool for investigating information processing within particular brain regions" (Chapin et al. 1999, p. 669), and that brain-machine systems "can become the core of a new experimental approach with which to investigate the operation of neural systems in behaving animals" (Nicolelis 2003, p. 417). What these researchers are alluding to is not only the mere possibility of using brain-machine interfaces to acquire and study neural signals. Their claim is that bionic systems composed of biological and artificial parts may play experimental roles in the research on human and animal behaviour, in the sense clarified above: observing behaviour of hybrid systems in controlled settings may provide empirical grounds to accept or refute theoretical models of animal and human behaviour. The following sections are devoted to clarifying what experimental procedures can be adopted and what are their conditions of validity, based on a few examples taken from the scientific literature. Notably, two broad classes of bionics-supported methodologies will be identified—called replacement and connection methodologies—which only partially conform to the synthetic method.

12.4.2 Replacement Methodology

Let M be a purely notional theoretical model of B referring to functional components b1, b2, and b3 (see Fig. 12.1 left and Table 12.1). Suppose also that M describes the putative behaviour of each component: for example, if M is couched in neuroscientific terms, suppose that component b1 is characterized in terms of a regularity r(b1) between some "input" and "output" neurons or neural populations. Testing whether M is a plausible model of B may require one to find out whether B does have a component exhibiting behaviour r(b1). In principle, bionic technologies may assist one in addressing a research question of this form according to the following strategy.

Indeed, one might build an artificial component a1 that *is known* to display behaviour r(b1), replace the target component b1 with a1, thus obtaining a bionic system H, and finally check whether H's behaviour matches the behaviour of B. In that case, one may conclude that b1 exhibits r(b1), thus corroborating one of the theoretical claims made in M. Indeed, if H's behaviour does not change after replacing b1 with an artificial component a1 displaying behaviour r(b1), then it is reasonable to conclude that b1 displays behaviour r(b1) too. If, on the other hand, H's behaviour changes as a result of the replacement, one might conclude that the behaviour of a1 does not match the behaviour of b1, and therefore that b1 does not exhibit r(b1). Note that in this methodology the target component—i.e., the component whose behaviour is under investigation—is replaced by an artificial component in the framework of a resulting hybrid system: for this reason, it will be called a *replacement* methodology. Note also that this methodology is somehow akin to the synthetic method discussed above, since theoretical conclusions about B are obtained by comparing the behaviours of B and H. However, in this case, the theoretical conclusion itself selectively concerns only one aspect of the theoretical

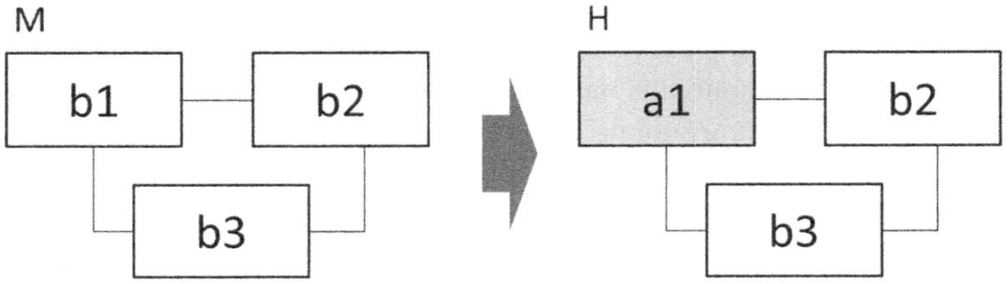

Fig. 12.1 Schema illustrating the relationship between a model M of the target system B (left) and a model H of the hybrid system (right) in the replacement strategy

Table 12.1 Labels used in the text and descriptions

Label	Description
B	A biological system
b1, b2, b3, ...	Functional components of biological system B
M	Theoretical model of B
r(b1)	Regularity characterizing component b1
a1, a2, a3, ...	Artificial components
r(a1)	Regularity characterizing component a1
H	Hybrid system
rs	The reticulo-spinal pathway in (Zelenin et al. 2000)
r(*rs*)	Regularity characterizing component *rs*
Vr	the vestibulo-reticular component in (Karniel et al. 2005)
A1, ... Ai, ..., An	in (Karniel et al. 2005), fully artificial systems which include computer simulations of different models of *vr*

model, and the system with which B's behaviour is compared is not fully artificial (as in the synthetic method).

This methodology is exemplified in a bionics-supported study on lamprey sensory-motor coordination (Zelenin et al. 2000). Lampreys maintain a stable roll position by moving tail, dorsal fin, and other body parts in response to external disturbances caused by water turbulence or other factors. A portion of the lamprey nervous system—called the reticulo-spinal pathway, *rs* from now on– is thought to play a crucial role in this behaviour. The goal of Zelenin and co-authors was to discover the behaviour of *rs*—or more precisely, to discover the relationship between the "input" neurons of *rs* (the reticular neurons) and the roll angles of the animal, which vary as a function of the activity of the "output" spinal neurons. The authors initially formulated a relatively simple hypothesis r(*rs*) whose content is not presented here, as it is irrelevant to understand the structure of the replacement method. To test this hypothesis, they built an electro-mechanical device whose input-output behaviour was known to be r(*rs*). They then functionally removed the reticulo-spinal component and replaced it with the electro-mechanical device: the artificial component picked up the activity of the reticular neurons and produced stabilization movements in line with the hypothesized regularity. Finally, Zelenin and colleagues experimentally tested whether the hybrid system exhibited stabilization abilities comparable to those of the intact system. This happened to be the case: the authors therefore concluded that the electro-mechanical device was a good substitute for the *rs* component—and, consequently, that the *rs* component actually exhibited the hypothesized input-output regularity r(*rs*).

Note that, to safely infer the behaviour of the replaced component from the behaviour of the hybrid system in line with the strategy outlined here, some auxiliary assumptions are required. First, one must exclude artefacts introduced by the brain-machine interface itself—for example, disturbances introduced by deteriorated electrodes. Second, one must ensure that the artificial device a1 produces the hypothesized regularity r(b1)—otherwise the experiment cannot shed light on whether b1 is governed by r(b1). Third, one must exclude significant plastic changes in the non-replaced portions of H (represented by b2 and b3 in Fig. 12.1). Adaptations of this kind occur frequently in bionic systems and are required for the proper functioning of prostheses or brain-controlled communication devices. However, they must be excluded in order to theorize about the behaviour of b1 based on the behaviour of H. Indeed, due to plastic adaptation processes, the hybrid system might end up displaying satisfactory stabilization behaviour, even if the artificial component a1 was not behaviourally equivalent to b1. In other terms, plastic processes might make the hybrid system successfully adapt to a "wrong" artificial component. In that case, the good behavioural performances of the hybrid system could not authorize one to conclude that a1 matches the behaviour of b1.

12.4.3 Connection Experiments

In replacement studies, the target component is replaced with an artificial device. Other bionics-supported experiments adopt a different approach: the target biological component is *connected* to, rather than replaced by, artificial components whose behaviour is known (such as a2 and a3 in Fig. 12.2). Then one observes H's behaviour and, also considering the known behaviour of a2 and a3, infers the behaviour of b1. This method is exemplified by a study on the lamprey nervous system reported in Karniel et al. (2005) and Mussa-Ivaldi et al. (2010). The goal was to discover the behaviour of the so-called vestibulo-reticular component—*vr* from now on—which, in the intact lamprey, receives vestibular input and modulates the activity of reti-

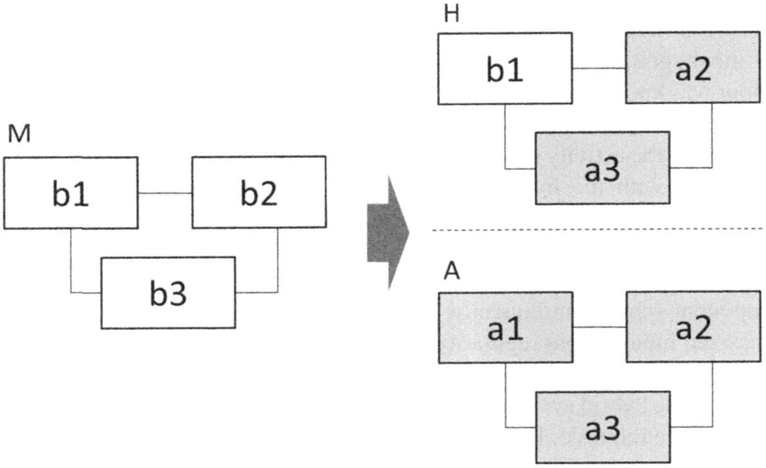

Fig. 12.2 Schema illustrating the relationship between a model M of the target system (left), a model H of the hybrid system (right, top), and a model A of a fully artificial simulation (right, bottom) in the connection strategy

culo-spinal neurons. To characterize the behaviour of *vr* (in terms of a relationship between input vestibular neurons and output reticular neurons) the authors extracted *vr* and placed it in the control loop of a robotic system. Light sensors mounted on the robot stimulated the input vestibular neurons, and the activity of the output reticular neurons was sent to the motor drivers of the robot. A hybrid system H, with biological component *vr* mediating between sensors and motors, was thus obtained. Note that the authors knew the behaviour of the artificial components of H, and for this reason were able to reflect on the target component's behaviour based on the overall behaviour of the system. For example, observing that the robot began to move along complex and curvilinear pathways when exposed to a single source of light, they concluded that the input-output relationship characterizing the vestibulo-reticular component could not be linear. So far, the methodology is substantially different from the synthetic method, as it does not involve any comparison between the behaviour of the target system and the behaviour of the bionic system.

Ideally, one may proceed by replacing b1 in H with an artificial component a1, thus obtaining a fully artificial system A. If H's behaviour turns out to be relevantly similar to A's behaviour, then one may conclude that b1's behaviour is relevantly similar to a1's behaviour. Given that the behaviour of a1 is known, one knows the behaviour of b1 as well. Note that this is a sort of replacement strategy which, unlike the example discussed in sect. 12.4.2, is applied to the bionic system and not to the target system. In the replacement strategy (see Fig. 12.1), one takes the target component out of the target biological system. In this case, one takes the target component out of the hybrid system, detaching it from the artificial components it had been connected to. Note also that this strategy partially conforms to the synthetic method, as the behaviour of a fully artificial system (A) is compared to the behaviour of a hybrid system (H). However, it does not fully conform to the replacement strategy in which, as discussed before, the behaviour of the hybrid system H is compared with the behaviour of the target system B (and not, as in this case, with the behaviour of the artificial system A).

This peculiar strategy was adopted in a particular phase of the lamprey study described here. After concluding that the input-output behaviour of *vr* could not be linear, the authors hypothesized a variety of possible internal models of it, differing from each other in the pattern of the connections among input and output neurons and, consequently, in the resulting input-output behaviour. They built computer simulations of each of these putative models. Then, the *vr* component was extracted from H and replaced with each of these "how-possibly" artificial vestibulo-reticular components in turn, thus obtaining an equal number of fully artificial systems A1, ..., An. Finally, the overt behaviour of each Ai system was compared with the overt behaviour of H. The idea was that the Ai system whose behaviour most closely matched H's behaviour would contain the artificial vestibulo-reticular component whose behaviour most closely matched *vr*'s behaviour. This experimental test, plus other theoretical considerations, led the authors to conclude that the target component's input-output behaviour was best captured by a mathematical model with recurrent ipsilateral connections—i.e., whose output depended not only on the instantaneous input, but also on the output produced immediately before by the same neuron. In this way, they developed a mathematical characterization of the input-output behaviour of *vr*.

One interesting feature of this strategy is that it may assist one in theorizing on the mechanisms governing plasticity in the target component. Recall that the authors had full control over the sensory inputs delivered to the target tissue. To explore plasticity, they reduced the gain of the left sensor, thus making the left input neuron significantly under-stimulated (with respect to the right sensor) in normal light conditions. The "left-blinded" hybrid system was left wandering around the experimental arena for 20 min. Then, the sensory gain was re-set at its previous value. A strong tendency to turn towards the right side of the workspace was observed thereafter. To explain this phenomenon, the authors trialled a series of alternative adjustments to the weights of the connections in the recurrent ipsilateral model in order to test which was most effective in reproducing the deteriorated behaviour observed in the left-blinded system. They eventually concluded that the tendency to turn to the right was caused by "a general reduction in the responsiveness of the reticular neurons on the right side, induced by a period of reduced contralateral input from the left vestibular afferents" (Karniel et al. 2005, p. S261).

One aspect of this example is worth noting. Recall that a "no-plasticity" constraint is required to infer the behaviour of the target biological component in replacement experiments (as discussed in the previous section): no plasticity process must occur in the non-replaced, thus *non-target* biological part of the hybrid system. In connection experiments, however, the target component is not replaced with, but connected to, artificial devices. Thus, the occurrence of plastic changes *in the target component* (as in the study described here) does not violate the "no-plasticity constraint" ruling out plastic changes *in the non-target biological components* of the bionic system. For other examples of connection experiments, see the study on thalamic circuitry described in Le Masson et al. (Le Masson et al. 2002) and the study on insect sensory-motor control described in Kanzaki et al. (Kanzaki et al. 2013).

12.5 Concluding Remarks

Automata have not only inspired mechanistic theories of life and behaviour. They have also served as experimental tools to test theoretical models of, and to predict, animal and human behaviour. Philosophers of science and technology have often reflected about the experimental roles played by robots in the cognitive and neural sciences, aiming to understand what kinds of theoretical conclusions can be obtained using robotic models, what experimental procedures can be adopted, what methodological complexities these procedures give rise to. Epistemological analyses of this kind may allow one to gain a broader understanding of the methods adopted in biological research, and of how inanimate models can be used to develop mechanistic explanations of the behaviour of living organisms. However, philosophical reflections about robotic models of animal and human behaviour have so far almost exclusively focused on the synthetic method—called here model-oriented strategy—involving fully artificial robots. Yet, especially in the last decades, robot-supported experiments which do not totally conform to the synthetic method have been reported in the scientific literature. They may diverge from the synthetic method because theoretical results are not inferred from comparison between artificial and biological behaviour, as in the predictive-oriented and in the connection strategy. Or, they may substantially conform to the synthetic method, but the system with which the target system is compared to a hybrid bionic system. This article has offered an overview of these experimental possibilities and reflected on the conditions under which they may provide genuine knowledge about the behaviour of biological systems.[1] It has suggested that the landscape of the possible experimental uses of robots in the life sciences is richer than what philosophers have envisaged so far.

And there are good reasons to believe that it is even richer, indeed. More specifically, bionics technologies enable methods which conform neither to the replacement nor to the connection strategy as presented here (widely discussed in Datteri 2017b). Moreover, in a growing number of studies, observation of the dynamics of interaction between robots and biological systems leads to theoretically interesting results on the mechanisms of social behaviour. These studies radically diverge from the synthetic method in many respects, including the fact that what is observed is not the behaviour of a robot (or hybrid system) in isolation, but the way biological systems interact with robots. A methodological analysis of these strategies may contribute to shedding further light on how automata can inform theorizing on animal and human behaviour in the contemporary life sciences.

References

Achinstein, P. (1965). Theoretical models. *The British Journal for the Philosophy of Science, XVI*(62), 102–120.

Braitenberg, V. (1986). *Vehicles. Experiments in synthetic psychology*. Cambridge: The MIT Press.

Brooks, R. (1999). *Cambrian intelligence: The early history of the new AI*. Cambridge: The MIT Press.

Chapin, J. K., Moxon, K. A., Nicolelis, M. A., & Markowitz, R. S. (1999). Real-time control of a robot arm using simultaneously recorded neurons in the motor cortex. *Nature Neuroscience, II*(7), 664–670.

Chirimuuta, M. (2013). Extending, changing, and explaining the brain. *Biology and Philosophy, XXVIII*(4), 613–638.

Cordeschi, R. (2002). *The discovery of the artificial. Behavior, mind and machines before and beyond cybernetics*. Dordrecht: Springer Netherlands.

Craver, C. F. (2010). Prosthetic models. *Philosophy of Science, LXXVII*(5), 840–851.

[1] For more thorough discussions of specific points raised here, see Craver (2010), Datteri (2009, 2017a), and Datteri and Schiaffonati (2019). For a specific discussion on the role of bionic systems in studying plasticity, see (Chirimuuta 2013) and (Datteri 2017b).

Datteri, E. (2009). Simulation experiments in bionics: A regulative methodological perspective. *Biology and Philosophy, XXIV*(3), 301–324.
Datteri, E. (2017a). Biorobotics. In L. Magnani & T. W. Bertolotti (Eds.), *Springer handbook of model-based science* (pp. 817–837). Heildelberg: Springer.
Datteri, E. (2017b). The epistemic value of brain–machine Systems for the Study of the brain. *Minds and Machines, XXIVV*(2), 287–313.
Datteri, E., & Schiaffonati, V. (2019). Robotic simulations, simulations of robots. *Minds and Machines, XXIX*(1), 109–125.
Datteri, E., & Tamburrini, G. (2007). Biorobotic experiments for the discovery of biological mechanisms. *Philosophy of Science, LXXIV*(3), 409–430.
Glennan, S., & Illari, P. (Eds.). (2018). *The Routledge handbook of mechanisms and mechanical philosophy*. New York: Routledge.
Grey Walter, W. (1950). An imitation of life. *Scientific American, CLXXXII*(5), 42–45.
Grasso, F. W., Consi, T. R., Mountain, D. C., & Atema, J. (2000). Biomimetic robot lobster performs chemo-orientation in turbulence using a pair of spatially separated sensors: Progress and challenges. *Robotics and Autonomous Systems, XXX*(1–2), 115–131.
Kanzaki, R., Minegishi, R., Namiki, S., & Ando, N. (2013). Insect-machine hybrid system for understanding and evaluating sensory-motor control by sex pheromone in Bombyx mori. *Journal of Comparative Physiology. A, Neuroethology, Sensory, Neural, and Behavioral Physiology, CXCIX*(11), 1037–1052.
Karniel, A., Kositsky, M., Fleming, K. M., Chiappalone, M., Sanguineti, V., Alford, S. T., & Mussa-Ivaldi, F. (2005). Computational analysis in vitro: Dynamics and plasticity of a neuro-robotic system. *Journal of Neural Engineering, II*(3), S250–S265.
Le Masson, G., Renaud-Le Masson, S., Debay, D., & Bal, T. (2002). Feedback inhibition controls spike transfer in hybrid thalamic circuits. *Nature, CDXVII*(6891), 854–858.
Long, J. (2012). *Darwin's devices. What evolving robots can teach us about the history of life and the future of technology*. New York: Basic Books.
Long, J. H., Schumacher, J., Livingston, N., & Kemp, M. (2006). Four flippers or two? Tetrapodal swimming with an aquatic robot. *Bioinspiration & Biomimetics, I*(1), 20–29.
Magnani, L., & Bertolotti, T. (Eds.). (2017). *Springer handbook of model-based science*. Cham: Springer International Publishing.
Mussa-Ivaldi, F. A., Alford, S. T., Chiappalone, M., Fadiga, L., Karniel, A., Kositsky, M., & Vato, A. (2010). New perspectives on the dialogue between brains and machines. *Frontiers in Neuroscience, IV*(1), 44–52.
Nicolelis, M. A. L. (2003). Brain-machine interfaces to restore motor function and probe neural circuits. *Nature Reviews Neuroscience, IV*(May), 417–422.
Pfeifer, R., & Bongard, J. (2006). *How the body shapes the way we think. A new view of intelligence*. Cambridge, MA: The MIT Press.
Riskin, J. (2016). *The restless clock: A history of the centuries-long argument over what makes living things tick*. Chicago, IL: University of Chicago Press.
Simon, H. A., & Newell, A. (1962). Computer simulation of human thinking and problem solving. *Monographs of the Society for Research in Child Development, XXVII*(2), 137–150.
Tamburrini, G., & Datteri, E. (2005). Machine experiments and theoretical modelling: From cybernetic methodology to neuro-robotics. *Minds and Machines, XV*(3–4), 335–358.
Truitt, E. R. (2016). *Medieval robots. Mechanism, magic, nature, and art*. Philadelphia, PA: University or Pennsylvania Press.
Webb, B. (2001). Can robots make good models of biological behaviour? *The Behavioral and Brain Sciences, XXIV*(6), 1033–1050.
Webb, B. (2008). *Using Robots to Understand Animal Behavior. Advances in the Study of Behavior, XXXVIII*, 1–58.

Weisberg, M. (2013). *Simulation and similarity. Using models to understand the world*. Oxford: Oxford University Press.
Winsberg, E. B. (2010). *Science in the age of computer simulation*. Chicago, IL: University of Chicago Press.
Zelenin, P., Deliagina, T., Grillner, S., & Orlovsky, G. (2000). Postural control in the lamprey: A study with a neuro-mechanical model. *Journal of Neurophysiology, LXXXIV*(6), 2880–2887.

Chapter 13
Knowledge as Duty

Morality in a Technological World

Lorenzo Magnani

Abstract Modern technology has brought about consequences of such magnitude that old policies and ethics can no longer contain them. Technological advances have made nature an object of human responsibility, and, as a result, if we are to restore and ensure her health, we must employ clever new approaches and rich, updated scientific and ethical knowledge. In our technological world, producing and applying such knowledge to real situations has become a duty, one that is just as important as making scientific advances.

Keywords Morality · Technology · Knowledge as duty · Moral mediators · Distributed morality

13.1 Introduction

In this article I will present some issues concerning philosophy and ethics of technology that are treated in my book *Morality in a Technological World. Knowledge as Duty*, published in 2007 by Cambridge University Press, also translated in Italian[1] In this book I have extendedly contended that anthropocentric ideas, like those that inform Kant's imperative, have made it difficult for people to acquire moral values usually associated with things and for things to attain moral worth traditionally

[1] First edition *Conoscenza come dovere. Moralità distribuita in un mondo tecnologico*, Pavia, Associated International Academic Publishers, 2005 (second Italian edition, Aracne, Rome, 2019).

L. Magnani (✉)
Computational Philosophy Laboratory, Università degli Studi di Pavia, Pavia, Italy
e-mail: lmagnani@unipv.it

S. Chiodo, V. Schiaffonati (eds.), *Italian Philosophy of Technology*, Philosophy of Engineering and Technology 35,
https://doi.org/10.1007/978-3-030-54522-2_13

reserved for people. We said that, in Kantian terms, people do not have to be "treated as means (and only as means)". I propose replacing that idea with a new one—respecting people as things in a positive sense. In this scenario, people are respected as "means" in a way that creates a virtuous circle, one in which positive moral aspects enjoyed by things can be used to reshape moral endowments attributed to things.

Attributing moral worth to things can be seen as a combination of the Kantian imperative and of John Stuart Mill's idea of freedom: "The only freedom which deserves the name, is that of pursuing our own good in our own way, so long as we do not attempt to deprive others of theirs, or impede their efforts to obtain it" (Mill 1966, p. 18). If, as Mill teaches, beings (or things, we now add) have the right to something, they are entitled not only to the goal itself but also to the unobstructed pursuit of it. When things also became regarded as entities with interests and rights of their own, the philosophical conceptual space of utilitarianism (animals suffer!) and the idea of environmental ecology were constructed. How did this happen?

A special kind of "things", animals, has always been used to lend some aspects of their own properties and functions to human beings, for instance in biomedical research. In this field animal models have served to induce certain conditions in animals for deriving conclusions about some conditions in human beings. The results are achieved by exploiting analogies (for instance rats and humans are alike in various ways) rather than disanalogies. This theme is very important in philosophy of science, because modeling is widespread in scientific practice. Many epistemological problems arise, like the quest of the qualities that make an animal model valid and appropriate (Shelley 2006). I contend that also in the area of ethics we have to look at the moral "models" that come from things, like animals and objects: as I already said, people can be respected as "means" in a way that creates a virtuous circle, one in which positive moral aspects enjoyed by things can be used to reshape moral endowments attributed to humans.

Perhaps the first "things" to gain new moral rights in Western culture were women, a change that was not universally welcomed. Indeed, the ideas propagated in this direction by Mary Wollstonecraft in her 1792 treatise *A Vindication of the Rights of Women* were initially considered absurd. This sort of ideological conflict has been played out again in the last few decades as animal rights advocates and environmental ethicists have waged a struggle similar to the one woman faced in the eighteenth century—that of redefining a means as an end. To achieve that goal, some intellectuals and activists have sought to reframe how various plants, animals, ecosystems—even the land itself—are valued so that they are regarded as "ends" and accorded the rights and protection that status entails.

13.1.1 Dehumanizing Humans

Reintegrating humanity into the natural world (Kirkman 2002), as the environmental philosophers do, means in some sense dehumanizing and renaturalizing humans. A tacit assumption could be that humans are capable of choices and actions, i.e., they are free (and consequently antinatural) and so can decide to abandon a previous state of harmony with nature. This anthropocentrism, when mixed with ignorance and arrogance, can affect environmental equilibria. It is well-known that, to contrast the consequent negative effects, environmentalism attributes to "things" certain intrinsic moral values (and legal rights too) previously held only by humans. At the same time, human beings have sought values and claimed characteristics that have traditionally been associated only with things, qualities like naturality and a "primitive" belonging to the whole of nature: human subjectivity and natural objectivity are both encompassed by the supposed "subjectivity" of an organic wholeness (Kirkman 2002, p. 33); in this way, the standard moral ontologies are modified. We already know freedom belongs to the realm of Kantian transcendence (Kant 1964), and so it is abstracted from the materiality of the biological determination of human beings. On the contrary, some biologists, who tend to see most human problems in terms of their science, sometimes maintain that freedom is a kind of biologically adaptive illusion shaped by evolution (Wilson 1998). From the environmentalist point of view, human freedom is just one aspect of the spontaneity of nature and, as we have already observed, ethics is a product of an evolution that favors human cooperation (Leopold 1998).

13.1.2 Reifying Humans and Humanizing Things

Once moral intrinsic values are attributed to external "things"[2] (and to some technological artifacts), for example, those "things" become limitations to human choices and create new problems. We have already described this idea as an extension of the Millian "freedom of pursuing our own good in our own way" without attempting "to deprive others of theirs, or impede their efforts to obtain it". By attributing intrinsic moral values to "external objects" as well as to other "people", we increase the number of entities we must respect when making choices, thereby limiting our freedom. Also the environmentalist modifications of ontologies (new values attributed to the environment) transform the usual ethical, legal, and political problems concerning regulations among human beings. Things (also some of the technological ones) are humanized, and humans, in turn, are reified as we make explicit how they affect and depend on "things". More comprehensive scientific and ethical knowledge is now at our disposal, which will allow us—humans—to make better decisions.

Humans and non-humans are deeply integrated, so much so that it becomes profoundly reductive to see people's problems only in terms of the selfish economical needs of global capitalism. While they hold various conceptions of nature and of their relationship with non-humans, all humans are involved in the natural world. This idea springs not just from the "metaphysical" datum that declares that we as

[2] Non technological things endowed with moral values are for example written laws, religious objects, objects we love, important artistic buildings, etc.

living bodies share the destiny of nature and of other things (necessity), and that as free spirits the realm of transcendence (freedom); it arises simply from the fact that, in ways that we are not yet able to appreciate and know, we are cognitively and morally integrated with other beings, and consequently, any changes we inflict on the world are also inflicted on us.

13.1.3 Human and Non-human Collectives

Humans and not-humans (also technological artifacts) are inextricably intertwined: "You are different with a gun in your hand; the gun is different with you holding it" (Latour 1999, p. 179). We are in some sense "folded" into non-humans, so that we delegate action to external things (objects, tools, technological artifacts) that in turn share our human existence with us. The idea of the "collective" expresses an exchange of human and non-human properties akin to what I have just described in the case of things in search of intrinsic values: "what the modernist science warriors see as a horror to be avoided at all costs—the mixing up of objectivity and subjectivity—is for us, on the contrary, the hall mark of a civilized life" (Latour 1999, p. 2000).

Many such examples are mentioned by Bruno Latour: using knowledge about non-humans to reconfigure people and, conversely, projecting on non-humans the properties and functions of humankind. When considered from the ethical perspective, the first case depicts our problem of respecting people as things, while the second depicts ideas which regard the moral representations of non-humans: "The new hybrid remains a non-human, but not only has it lost its material and objective character, it has acquired properties of citizenship" (Latour 1999, p. 202). Of course, the non-moral case of endowing non-human entities with speech, intelligence, and other human properties—technological things from classical media to computational tools, from paintings to artificial intelligence, from simple tools like a hammer to sophisticated machines—is related to this movement. So, too, are agriculture and the domestication of non-human animals, which involves their socialization and reeducation.

In turn, external things (electrical, transportation, and telecommunication industries, for example) have constructed new social frameworks for people, and so in the case of the many new roles delineated by factories, machines, and institutions in establishing new constraints in managing humans and in stabilizing new types of human negotiations: "It was from techniques, that is, the ability to nest several subprograms, that we learned what it means to subsist and expand, to accept a role and discharge a function" (Latour 1999, p. 209). Tools (and so various technological artifacts), that have always played the role of human prostheses, become integrated into our bodies as we use them in a kind of anthropological transformation of both the individual and the collectives. This mixture between human and non-human is also expressed in human bodies that are increasingly shaped and integrated by sociotechnical negotiations and artifacts.

The cyclical process of transferring qualities between humans and non-humans is, of course, an inextricable part of our evolution, and consequently it requires ongoing negotiations and a continual redrawing of the lines between the two kinds of entities.

13.1.4 Cognitive Things/Cognitive Beings and Technology

Tools and utensils used throughout human history were mainly an extension of men's and women's bodies; in this sense they did not possess an independent existence and seemed to be in harmony with the environment. More recent technological artifacts, on the other hand, have independent sources of power and exist separately from the user. They have established a kind of third estate midway between nature and humane arts: "While many of the boasted achievements of industrialism are merely rubbish, and while many of the goods produced by the machine are fraudulent and evanescent, its esthetic, its logic, and its factual technique remain a durable contribution: they are among man's supreme conquests" (Mumford 1961, p. 5).

Technology represents intelligence systematically applied to the body. It constitutes a kind of prosthesis that amplifies the body and transcends its limits, compensating for its fragility and vulnerability. Because of industrial technology, the human body is capable of more than ever before, and society's unprecedented production capacity exceeds anything thought possible in the past. Usually it is said that this process has diminished the importance and diluted the talent of the worker. It is also said that technologies do not merely add something new to an environment; instead, they change the whole environment itself—ecologically, structurally, or both (Strong 2000). By anthropomorphizing things, computers, for example, we begin to devalue and disempower people while attributing too much power and value to computer technologies. From paper and pencil to computers, we have invented skillful "cognitive things". Externalizing human cognitive qualities onto machines and technological artifacts is followed by the internalizing of machine's cognitive qualities onto humans. Thinking with a computer differs from thinking with paper and pencil; the computer creates a new environment in which the mind breathes a different atmosphere. Is it an information-rich world or an information-polluted world?

The new metaphors can involve a kind of "identity crisis": "It makes a difference whether we speak of a computer as having a 'memory' or a 'data-storage capacity'. As a result of this crisis we have less sense of what is human, less sense of humans as distinct from machines, more sense of powerful machines and frail humans" (Gozzi 2001, p. 148). The computer is a brain, advanced systems show an artificial intelligence, machines have their own language, the robots have kinesthetic abilities. These technologies are the result of the passage from the idea of technology, as conceived like a "science or systematic knowledge of human arts of industrial arts" to the one exclusively in terms of objective artifacts: "We speak of things as being high-tech, not people" (Gozzi 2001, p. 148). In some sense, regarding technology this way renders it autonomous and removes it from its dependence on human beings. I agree with Gozzi: "This issue is not trivial, for it involves our definition of ourselves as human beings. Which is somewhat uncertain at present" (Gozzi 2001, p. 148). Humans beings can seem frail and contradictory compared to the strong, stable, and reliable machines that surround us. Of course the case of humanoids, cyborgs, and bionic body parts complicate the issue.

Qualities transferred from things to people and vice versa often carry troubling overtones: the idea of the person as a machine, which dates back to the nineteenth century, has a negative connotation, as does the phrase "organism's program". It is also said, in an ominous way, that one "deprograms" people when dissuading them from certain convictions—political or religious beliefs, for example: "Are people machines? If so, they are clearly inferior to the faster, bigger models; and they deserve to be made obsolete. Are biological processes determined by laws of chemistry and physics? If so why bother giving people all those troublesome rights and freedoms, which are illusory anyway?" (Gozzi 2001, p. 148).

Humans beings are reduced to things in all these cases, making it is easier to treat people as means. Teasing out the meaning and significance of these complex issues is our duty, especially when dealing with problems in collective settings like work, school, and politics as well as in family arenas, where sex, children, relationships, and, as we have seen, reproduction can create challenges.[3]

In turn, I think the concept of "respecting people as things" provides an ethical framework through which to analyze the condition of modern people, who, as increasingly commodified beings, are becoming more and more thing-like anyway.

13.2 Moral Mediators, Moral Reasoning, and Technological Artifacts

In my book *Morality in a Technological World. Knowledge as Duty* (Magnani 2007) I extendedly considered numerous ethical issues related to technology: ecology, biotechnology, the hybridization of human beings, cyberprivacy, bad faith, globalization, and the unethical effects of external systems and technologies in general. Each of these discussions underscored the importance of producing and exploiting appropriate ethical knowledge and reinforced my argument that knowledge is a duty. If, as I contend, new ethical, scientific, and other kinds of understandings must be developed and implemented, then cognitive concerns also become fundamentally important. Indeed, in the last two chapters of the book I closely examined the cognitive aspects of moral mediators and of other methodological problems related to ethical reasoning and moral deliberation, as I will briefly describe in this section.

Following seminal studies in philosophy of science related to cognitive science (Magnani and Nersessian 2002; Magnani and Nersessian 2002) it is well-known

[3] I addressed this duty in Chap. 4 of my book, contending that, if fulfilled, it can enhance the freedom and responsibility of human beings, as described in detail in Chap. 5.

that not only are ethical knowledge and reasoning expressed at the verbal/propositional level, they can also involve model-based (visual, for example) and manipulative/"through doing" aspects: for example, an important component of ethics is imagination, which is—together with analogy, visualization, simulation, the thought experiment, and so on—a form of model-based reasoning. Creativity is also important, for through it human beings expand knowledge and create new perspectives. To explain morality "through doing", I have illustrated in my book manipulative ethical reasoning using a list of invariant behaviors that I call "moral templates", which represent embodied patterns of possible moral behavior, either preexisting or newly created in people's mind-body systems, that enable a kind of moral "doing". I also think it is useful to cognitively compare moral deliberation to diagnosis, a strategy that reveals the logical details of the intrinsic incompleteness of knowledge in ethical inferences.

Using a cognitive and epistemological approach to the concept of abduction and model-based reasoning, as I proposed in my research (Magnani 2009, 2017),[4] produces an important and valuable side effect: an integrated view that forms a unique framework through which to study the multiple aspects of moral reasoning, including those that are verbal/propositional, model-based, distributed ("moral mediators"), and embodied ("templates of moral doing").

I have said that only human acts of cognition can add worth to or subtract value from an entity, and that revealing the similarities between people and things can help us to attribute to human beings the kind of worth that is now held by many highly valued nonhuman things. This process suggests a new perspective on ethical thinking: indeed, these objects and structures can mediate moral ideas and recalibrate the value of human beings by playing the role of what I call *moral mediators*.

What exactly is a moral mediator? I derived the concept of the moral mediator from that of the epistemic mediator, which I introduced in my previous research on abduction and creative and explanatory reasoning. First of all, moral mediators can extend value from already-prized things to human beings, as well as to other nonhuman things and even to "non-things" like future people and animals. We are surrounded by human-made and artificial entities, whether they are concrete objects like a hammer or a PC or abstractions like an institution or society; all of these things have the potential to serve as moral mediators. For this reason, I say that it is critically important for current ethics to address not only the relationships among human beings, but also those between human and nonhuman entities. Moreover, by exploiting the concepts of "thinking through doing" we can see that a considerable part of moral action is performed in a tacit way, so to say, "through doing". Part of this "doing" can be considered a manipulation of the external world in order to build various moral mediators that function as enormous new sources of ethical information and knowledge. As I said, I call these schemes of action "templates of moral doing".

[4]Abductive cognition refers to all kinds of inferences able to produce plausible hypotheses, such as, for example, in the case of diagnostic reasoning or in scientific discovery.

In the cases just mentioned, moral mediators are purposefully constructed to achieve particular ethical effects, but other aspects and cognitive roles of moral mediators are equally important: moral mediators are also beings, entities, objects, and structures that objectively, even beyond human beings' intentions, carry possible ethical or unethical consequences. External moral mediators function as components of a memory system that crosses the boundary between person and environment. For instance, when a society moves an abused child into a foster home, it is seeking both to protect her and to reconfigure her social relationships; in this case, the new setting functions as a moral mediator that changes how she relates to the world—it can supply her with new emotions that bring positive moral and psychological effects and help her gain new perspectives on her past abuse and on adults in general. I depict these processes as "model-based" inferences, and indeed one way moral mediators transform moral tasks by promoting further moral inferences in agents at the level of model-based abduction (Magnani 2009). I use the term "model-based reasoning" to mean the constructing and manipulating of certain representations, not mainly sentential and/or formal, but mental and/or related to external mediators: obvious examples of model-based inferences include building and using visual representations, conducting thought experiments, and engaging in analogical reasoning. In this light, an emotional feeling also can be interpreted as a kind of model-based cognition. Of course, abductive reasoning—the process of reasoning to hypotheses—can be performed in a model-based way, either internally or with the help of external mediators.

Moreover, I can use manipulation to alter my bodily experience of pain; I can, for example, follow the behavior template "control of sense data" described below in this article, during which I might shift—often unconsciously—the position of my body. Through manipulation I can also change my body's relationships with other humans and nonhumans experiencing distress, as did Mother Theresa, whose rich, personal moral feeling and consideration of pain were certainly shaped by her physical proximity to starving and miserable people and by her manipulation of their bodies. In many people, moral training is often related to the spontaneous (and sometimes fortuitous) manipulation of both sense data and their own bodies, for these actions can build morality immediately and nonreflectively "through doing".

Technological artifacts serve as moral mediators in many situations, as in the case of certain machines that affect privacy. It is well-known that the internet mediates human interaction in a much more profound way than do traditional forms of communication like paper, the telephone, and mass media, even going so far as to record interactions in many situations. The problem is that because the internet mediates human identity, it has the power to affect human freedom. Thanks to the internet, our identities today largely consist of an externally stored quantity of data, information, images, and texts that concern us as individuals, and the result is a "cyborg" of both flesh and electronic data that identifies us. I contend that this complex new "information being" depicts new ontologies that in turn involve new moral problems. We can no longer apply old moral rules and old-fashioned arguments to beings that are simultaneously biological and virtual, situated in a three-dimensional local space and yet "globally omnipresent" as information packets. Our cybernetic locations are no longer simple to define, and increasing tele-presence technologies will exacerbate this effect, giving external, non-biological resources even greater powers to mediate ethical endowments such as those related to our sense of who and what we are and what we can do. These and other effects of the internet—almost all of which were unanticipated—are powerful motivators of our duty to construct new knowledge.

We have also to add that already in the nineties Maner underlined the so-called "uniqueness debate", which emphasizes the special character of computer ethics: "I have tried to show that there are issues and problems that are unique to computer ethics. For all of these issues, there was an essential involvement of computing technology. Except for this technology, these issues would not have arisen, or would not have arisen in their highly altered form. The failure to find satisfactory non-computer analogies testifies to the uniqueness of these issues. The lack of an adequate analogy, in turn, has interesting moral consequences. Normally, when we confront unfamiliar ethical problems, we use analogies to build conceptual bridges to similar situations we have encountered in the past. Then we try to transfer moral intuitions across the bridge, from the analog case to our current situation. Lack of an effective analogy forces us to discover new moral values, formulate new moral principles, develop new policies, and find new ways to think about the issues presented to us" (Maner 1996, p. 152, see also Johnson 1994).

I believe that in the context of this abstract but ubiquitous technological presence, certain moral approaches that ethics has traditionally tended to disparage are worth a new look. Taking care of both people and external things through personal, particular acts—a moral orientation often associated with women—rather than relating to others through an impersonal, general concern about humanity has a new appeal. The ethics of care does not consider the abstract "obligation" as essential; moreover, because it does not require that we impartially promote the interests of everyone alike, it allows us to focus on those who most need assistance.

In short, a considerable part of morality occurs in an implicit way, so to say, "through doing", and part of this "doing" features manipulating the external world in order to build various external "moral mediators" (also technological) that can provide vast amounts of new information and knowledge, transform ethical features and effects, and sometimes, of course, generate unethical outcomes. To further deepen basic aspects which refer to the role of moral mediators in philosophy and ethics of technology it is necessary to illustrate some other fundamental issues.

13.2.1 Hybrid People: Respecting Things as People

In chapters four and five of (Magnani 2007), I try to outline a new ethical framework that makes problems like identity, freedom, responsibility, and ownership of our own destinies more meaningful. In this subsection, however, let us reconsider the consequences of my motto "respecting people as things", when applied to the idea of hybrid people, as people which are strictly intertwined with technological artifacts that play the role of cognitive and moral mediators.

Following Andy Clark's. conclusions—illustrated in *Natural-Born Cyborgs: Minds, Technologies and the Future of Human Intelligence* (2003)—on the relationships between humans and technology, we all are "constitutively" natural-born cyborgs—that is, biotechnologically hybrid minds. Less and less are our minds considered to be in our heads: human beings have solved their problems of survival and reproduction by "distributing" cognitive functions to external non-biological sources, props, and aids. Our biological brains have delegated to external tools many activities that involve complex planning and elaborate assessments of consequences (Clark 2003, p. 5). A simple example might be how the brain, when faced with multiplying large numbers, learns to act in concert with pen and paper, storing part of the process and the results outside itself. The same occurred when Greek geometers discovered new properties and theorems of geometry: they manipulated external diagrams to establish a kind of continuous cognitive negotiation with a suitable external support (like sand or a blackboard), to gain new important information and heuristic suggestions.[5] The use of external tools and artifacts is very common: cognitive skills and performances are so widespread that they become invisible, thus giving birth to something I have called "tacit templates" of behavior that blend "internal" and " external" cognitive aspects.[6]

New technologies will facilitate this process in a new way: on a daily basis, people are linked to non-biological, more-or-less intelligent machines and tools like cell phones, laptops, and medical prosthetics. Consequently, it becomes harder and harder to say where the external world stops, and the person begins. Clark contends that this line between biological self and technological world has always been flexible and that this fact has to be acknowledged both from the epistemological and the ontological points of view. Thus, the study of the new anthropology of hybrid people becomes important, and I would add that it is also critical for us to delineate and articulate the related ethical issues. Some moral considerations are mentioned in the last chapter of Clark's book, in which he addresses important issues such as inequality, intrusion, uncontrollability, overload, alienation, narrowing, deceit, degradation, and disembodiment—topics that are especially compelling given recent electronic and biotechnological transformations. Nevertheless, Clark's approach does not shed sufficient light on basic ethical problems related to identity, responsibility, freedom, and control of one's destiny, problems that accompany technological transformations. He clearly acknowledges such issues, but only in a minimal and general way: "Our redesigned minds will be distinguished by a better and more sensitive understanding of the self, of control, of the importance of the body, and of the systemic tentacles that bind brain, body, and technology into a single adaptive unit. This potential, I believe, far, far overweighs the attendant threats of desensitization, overload, and confusion ... Deceit, misinformation, truth, exploration, and personal

[5] I have devoted part of my research to analyze the role of diagrams in mathematical thinking and geometrical discovery (Magnani 2001, 2002).

[6] Tacit templates of moral behavior in relation to moral mediators in the sixth chapter of (Magnani 2007). article. Their epistemological counterpart is illustrated in second chapter of (Magnani 2009).

reinvention: Internet provides for them all. As always, it is up to us, as scientists and as citizens, to guard against the worst and to create the culture and conditions to favor the best" (Clark 2003, p. 170, p. 187).

I think these problems are more complicated and teasing out their philosophical features will require deeper analyses. What new knowledge must we build to meet the challenges of living as hybrid people? I certainly share Clark's enthusiasm in philosophically acknowledging our status as "cyborgs", but I would like to go further, to do more than just peer through the window of his book at the many cyberartifacts that render human creatures the consumers-cyborgs we are.[7]

Our bodies and our "selves" are materially and cognitively "extended", meshed, that is, with external artifacts and objects, and this fact sets the stage for a variety of new moral questions. For example, because so many aspects of human beings are now simulated in or replaced by things in an external environment, new ontologies can be constituted—and Clark would agree with me. Pieces of information that can be carried in any physical medium are called "memes" by Richard Dawkins (1989). They can "stay" in human brains or jump from brain to brain to objects, becoming configurations of artificial things that express meaning, like words written on a blackboard or data stored on a SSD drive, icons and diagrams on a newspaper, configurations of external things that express meaning like an obligatory route. They can also exist in natural objects endowed with informative significance—stars, for example, which offer navigational guidance. In my perspective the externalization of these chunks of information is described in the light of the cognitive delegation human beings concentrate in material objects and structures, as when one jots down a phone number on the back of an envelope.

Beyond the supports of paper, telephone, and media, many human interactions are strongly mediated (and potentially recorded) through the Internet. What about the concept of identity, so connected to the concept of freedom? At present identity must be considered in a broad sense: the externally stored amount of data, information, images, and texts that concern us as individuals is enormous. This storage of information creates for each person a kind of external "data shadow" that, together with the biological body, forms a "cyborg" of both flesh and electronic data that identifies us or potentially identifies us. I contend that this complex new "information being" depicts new ontologies that in turn involve new moral problems. We can no longer apply old moral rules and old-fashioned arguments to beings that are at the same time biological (concrete) and virtual, situated in a three-dimensional local space but potentially "globally omnipresent" as information-packets. For instance, where we are located cybernetically is no longer simple to define, and the increase in telepresence technologies will further affect this point. It becomes clear that external, non-biological resources contribute to our variable sense of who and what we are and what we can do.

[7] Below I will address some details about the role of this kind of cognitive delegation from an ethical perspective.

13.2.2 Moral Mediators and Technology

We can say that a significant portion of human cognitive manipulations of the external world is also devoted to building a vast new source of information and knowledge: external moral mediators. In sum, moral mediators represent a kind of redistribution of the moral effort through managing objects and information in such a way that we can overcome the poverty and the unsatisfactory character of the moral options immediately represented or found internally (for example principles, prototypes, etc.). I also think that the analysis of moral mediators can help accounting for the mechanisms of the spectacle of global moral actions and collective responsibilities originating from the intertwining between numerous agents at local level.

We have already established that the monumental moral challenges created by technology are making it necessary for us to think about the world in different ways, and I do not believe that concepts like the moral agent/moral patient paradigm are sufficient to give us the tools we need to achieve this goal. I contend that framing new moral problems by employing the idea of moral mediators is a much more fruitful way to proceed.

Again, much of the behavior we conduct through learned habits is devoted to building vast new sources of information and knowledge: external moral mediators. I repeat, take the example of those human beings who create, in a tacit and embodied way, foster homes, which play a moral role: they facilitate a foster child's recovery as they allow him or her to rebuild moral perceptions that were damaged by previous abuse. Here, the artifact (the foster home, partially technological) is an example of a moral mediator in the sense that it mediates—objectively, "out there" in an external structure—positive moral effects. Other moral mediators, however, are built by human collectives in a more conscious way, as in the case of some objectified rules and principles created with a particular goal in mind or already having established ways of producing moral effects (religious rites, for examples). Many complicated external moral mediators can also redistribute moral effort: they allow us to manipulate objects and information in a way that helps us to overcome the paucity of internal moral options (principles and prototypes, etc.) currently available to us. I also think that moral mediators can help to explain the macroscopic and growing problem of global moral actions and collective responsibilities which result from the "invisible hand" of systemic interactions among several agents at the local level.

Using moral mediators is more than just a way to move the world toward desirable goals: it is an action that can play a moral role and therefore warrants moral consideration. We have said that when people do not have adequate information or lack the capacity to act morally upon the world, they can restructure their worlds in order to simplify and solve moral tasks. Moral mediators are also used to reveal latent constraints in the human/environment system, and these discoveries grant us precious new ethical information. Imagine, for instance, a wife whose work requires long hours away from her husband, and whose frequent absences cause conflict in their relationship. To improve their marriage, she restructures her life so that she can spend more quality time with her spouse, an action that can cause variables affected by "unexpected" and "positive" events in the relationship to covary with informative, sentimental, sexual, emotional, and, generally speaking, bodily variables. Before the couple adopted a reconfigured "social" order—that is, increased their time together—there was no discernible link between these hidden and overt variables; a new arrangement has the power to reveal important new "information", which, in our example, might come from a revitalized sex life, surprisingly similar emotional concerns, or a previously unrecognized intellectual like-mindedness.

In the previous sections we considered the problem of "respecting people as things" as we discussed the potential of external "natural" objects to create new ethical knowledge. Endangered species can serve as moral mediators, for example, when people define themselves as "endangered", for such comparisons provide us with new insights into other human beings' self-perceptions. In fact, many external things (and of course the technological ones) that have traditionally been considered morally inert can be seen as or transformed into moral mediators. For example, we can use animals, the earth, or cultural entities to identify previously unrecognized moral features of human beings, and we can employ external "tools" like writing, narrative, ritual, and, as we saw in the earlier examples of foster homes and group homes, technological tools such as computational ones, institutions to reconfigure unsatisfactory social orders. Hence, not all moral tools are merely mental—many are shared and distributed in external objects and structures that function as objectified ethical devices.

While almost any sort of entity can help to mediate our moral outlook, certain technological artifacts can be considered *über*-moral mediators—those equipped with artificial intelligence and the ability to be directly engaged in ethical reasoning and behavior. We must recognize the ethical ramifications not only of using such machines, but also of allowing them ethical autonomy *vis-à-vis* human users as well as other devices; in the process, we must assuage human fears about artificial intelligence. Developing ways to deal with artificial intelligence is a new field of research—called machine ethics (Anderson and Anderson 2011)—that involves many interesting topics: improving the interaction between artificial and natural intelligence systems by adding an ethical dimension to technological devices; using ethical strategies to enhance machine-to-machine communication and cooperation; developing systems that provide expert ethical guidance; establishing decision-making procedures for ethical theories with multiple *prima facie* duties that present conflicting perspectives; and assessing the impact of machine ethics on society.

External moral mediators of all kinds can function as components of a memory system that crosses the boundary between person and environment. For example, they transform the tasks involved in simple manipulations that promote further moral inferences at the level of model-based abductive hypothetical reasoning. In the earlier case of the wife seeking moral protection of her marriage, she transforms it by manipulating her behavior so as to increase the quality time spent with her husband, and thus discovers new information that allows her to abduce (that is to make appropriate hypotheses concerning…) new internal model-based ideas and/or her feelings—new motivating images, for example, or constructive emotions—

about her husband and/or her marriage. When the everyday life of a previously abused child is manipulated by placing her with a foster family, for instance, the new setting is a moral mediator that can help her abduce new model-based internal experiences, images, emotions, or analogies through which she may be able to recalibrate her conceptions of adults, her past, and of abuse in general.

Actions executed through tacit templates of behavior can even enhance one's level of physical sensitivity: I can alter my bodily experience of pain by following the previously mentioned control of sense data template that is, by unconsciously modifying the experience of my body and changing its relationships with humans or nonhumans in distress, I may, for instance, create new, empathic moral ways to help other beings. As I have already said, Mother Theresa's rich moral feeling and consideration of pain had certainly been shaped by her proximity to starving and miserable people and by her manipulation of both her own and their bodies. In many people, moral training is often related, as I have already observed above, to these kinds of spontaneous (and sometimes "lucky") manipulations of their own bodies and sense data so that they build morality immediately and nonreflectively "through doing."

Throughout history, women have traditionally been thought to place more value on personal relationships than men do, and they are often regarded as more adept in situations requiring intimacy and caring. It would seem that women's basic moral orientation emphasizes taking care of both people and external things through personal, particular acts rather than relating to others through an abstract, general concern about humanity. The ethics of care[8] does not consider the abstract "obligation" to be essential; moreover, it does not require that we impartially promote the interests of everyone alike. Rather, it focuses on small-scale relationships with people and external objects, so that, for example, it is important not to "think" of helping disadvantaged children all over the world (as men tend to aim at doing) but to "do" so when called to do so, everywhere.

My philosophical and cognitive approach to moral model-based thinking and to morality "through doing" does not mean that this so-called female attitude, being more closely related to emotion, should be considered less deontological or less rational and therefore a lower form of moral expression. I contend that many of us can become more intuitive, loving parents and, in certain situations, learn to privilege the "taking care" of our children by educating our feelings—maybe by heeding "Kantian" rules.[9] The route from reason to feeling (and, of course, from feeling to reason) is continuous in ethics. Many people are suspicious of moral emotional evaluations because emotions are vulnerable to personal and contextual factors. Nevertheless, there are moral circumstances that require at least partially emotional evaluations, which become particularly useful when combined with intellectual (Kantian) aspects of morality.

---

[8] The ethics of care is an ethical theory that considers moral action centered on interpersonal relationships and care or benevolence as a virtue. It is one of the main aspects of normative ethical theories that were developed by feminists in the second half of the twentieth century (Urban Walker 1996).

[9] The role of the ethics of care in bioethics is illustrated in (Carse 1999).

Consequently, "taking care"[10] is an important way to look at people and objects (and technological tools), and, as a form of morality accomplished "through doing", it achieves status as a fundamental kind of moral inference and knowledge. Respecting people as things is a natural extension of the ethics of care; a person who treats "nonhuman" household objects with solicitude, for example, is more likely to be seen as someone who will treat human beings in a similarly conscientious fashion. Consequently, using this cognitive concept, even a lowly kitchen utensil can be considered a moral mediator.

13.3 Knowledge as Duty

The title of this article already indicated the importance I attribute to the role of knowledge in our technological era: let try to better clarify the meaning of the motto "knowledge as duty". In our technological world, it has become critically important for us to produce and apply ethical knowledge that keeps pace with the rapid changes around us. We are no less obligated to pursue this knowledge than we are to seek scientific advances; indeed, to neglect the ethical dimension of modern technology is to court disaster. Recent advances have brought about consequences of such magnitude that old policies and ethics can no longer contain them, and we must be willing to approach problems in wholly new ways. Our technology has, for example, turned nature into an object of human responsibility, and if we are to restore and ensure her health, we must employ clever new approaches and rich, updated ethical knowledge. The scope and impact of our current technological abilities have handed human beings the responsibility for, say, "nature" and "the future", which were previously left to God or to fate. Consequently, I patently declare my hope for knowledge that maintains and enhances our endowments of intentionality, consciousness, and free-will choices; that strengthens our ability to undertake responsible action; and that preserves the ownership of our own futures.

Indeed, basic aspects of human dignity are constantly jeopardized not only by human mistakes and wrongdoing but also by technological products. Constant challenges also come from natural events and transformations, both ordinary ones, like the birth of one's first child, and extraordinary ones, like epidemics, tsunamis, and hurricanes. I think that preserving and improving the present aspects and characteristics of human beings depends on their own choices about knowledge and morality, and I believe strongly that knowledge is a primary duty that must receive much greater emphasis than ever before and that the knowledge we create must be commensurate with the causal scale of our action. I propose that one way to achieve this and other goals is by accepting "knowledge as duty" and by using disciplines like ethics, epistemology, and cognitive science to rethink and retool research on the philosophy of technology.

[10] On ethics of care and related problems see above Sect. 13.2.2.

13.4 Conclusion

The technological advances of contemporary society have outpaced our moral understanding of the problems that they create. How will we deal with profound ecological changes, human cloning, hybrid people, and eroding cyberprivacy, just to name a few issues? I fundamentally argue that existing moral constructs often cannot be applied to new technology. I have proposed an entirely new philosophical and ethical approach, one that blends epistemology with cognitive science. The resulting moral strategy promises new dignity for overlooked populations, who live in complicated technological environments, both of today and of the future.

Acknowledgements This article is excerpted from L. Magnani (2007), *Morality in a Technological World. Knowledge as Duty*, Cambridge, with permission from ©2007 Cambridge University Press.

References

Anderson, M., & Anderson, S. (Eds.). (2011). *Machine ethics*. Cambridge: Cambridge University Press.

Carse, A. L. (1999). Facing up to moral perils. The virtue of care in bioethics (1996). In H. J. Robinson, R. M. Berry, & K. McDonnell (Eds.), *A health law reader. An interdisciplinary approach* (pp. 51–59). Durham: Carolina Academic Press.

Clark, A. (2003). *Natural-born cyborgs. Minds, technologies, and the future of human intelligence*. Oxford-New York: Oxford University Press.

Dawkins, R. (1989). *The selfish gene (1976)*. Oxford: Oxford University Press.

Gozzi, R. (2001). Computers and the human identity crisis (1993). In D. Micah Hester & P. J. Ford (Eds.), *Computers and ethics in the Cyberage* (pp. 147–152). Upper Saddle River: Prentice-Hall.

Johnson, D. G. (1994). *Computer ethics* (2nd ed.). Englewood Cliffs, NJ: Prentice Hall.

Kant, I. (1964). *Groundwork of the metaphysics of morals (reprint of the 1956 edition, edited and translated by H.J. Paton. London: Hutchinson & co., third edition; originally published 1785)*. New York: Harper & Row.

Kirkman, R. (2002). Through the looking glass. Environmentalism and the problem of freedom. *The Journal of Value Inquiry, 36/1*, 27–41.

Latour, B. (1999). *Pandora's hope*. Cambridge: Harvard University Press.

Leopold, A. (1998). The land ethic (1949). In *A sand county almanac. And sketches here and there* (p. 1966). Oxford: Oxford University Press.

Magnani, L. (2001). *Philosophy and geometry. Theoretical and historical issues*. Dordrecht: Kluwer Academic.

Magnani, L. (2002). Epistemic mediators and model-based discovery in science. In L. Magnani & N. J. Nersessian (Eds.), *Model-based reasoning. Scientific discovery, technology, values*. (pp. 305–329). New York: Kluwer Academic/Plenum Publishers.

Magnani, L. (2007). *Morality in a technological world. Knowledge as duty*. Cambridge: Cambridge University Press (Italian edition *Conoscenza come dovere. Moralità distribuita in un mondo tecnologico*. Pavia: Associated International Academic Publishers, 2005. Second Italian edition, Roma: Aracne, 2019).

Magnani, L. (2009). *Abductive cognition. The epistemological and eco-cognitive dimensions of hypothetical reasoning*. Berlin-Heidelberg: Springer.

Magnani, L. (2017). *The abductive structure of scientific creativity. An essay on the ecology of cognition*. Cham: Springer.

Magnani, L., & Nersessian, N. J. (Eds.). (2002). *Model-based reasoning. Scientific discovery, technology, values*. New York: Kluwer Academic/Plenum Publishers.

Magnani, L., Nersessian, N. J., & Pizzi, C. (Eds.). (2002). *Logical and computational aspects of model-based reasoning*. Dordrecht: Kluwer Academic Publishers.

Maner, W. (1996). Science and engineering ethics. In T. Bynum & S. Rogerson (Eds.), *Unique ethical problems in information technology* (Vol. 2, pp. 137–154). (Special Issue: Global Information Ethics).

Mill, J. S. (1966). *On liberty (1859). On liberty, representative government, the subjection of women* (pp. 5–141). Oxford: Oxford University Press.

Mumford, L. (1961). Assimilation of the machine. New cultural values. In D. Micah Hester & P. J. Ford (Eds.), *Computers and ethics in the Cyberage* (pp. 3–7). Upper Saddle River: Prentice-Hall.

Shelley, C. (2006). Analogical reasoning with animal models in biochemical research. In L. Magnani (Ed.), *Model-based reasoning in science and engineering. Cognitive science, epistemology, logic* (pp. 203–213). London: College Publications.

Strong, D. (2000). Technological subversion. In M. E. Winston & R. D. Edelbach (Eds.), *Society, ethics, and technology* (pp. 148–159). Wadsworth/Thomson Learning: Belmont.

Urban Walker, M. (1996). Feminist skepticism, authority and transparency. In W. Sinnott-Armstrong & M. Timmons (Eds.), *Moral knowledge? New readings in moral epistemology* (pp. 267–292). Oxford: Oxford University Press.

Wilson, E. O. (1998). *Consilience. The unity of knowledge*. New York: Knopf.

Part IV
Aesthetics

Chapter 14
Art Inside Technology

Mario Costa

Abstract First, technologies (photography, tape recorder, telephone, radio, television) and, afterwards, neo-technologies (total digitalisation, Internet, mobile phones) have created the conditions for overcoming the arts towards what the author calls "technological sublime". Within this new dimension, not only the category of art reaches its own decline, but the very aesthetic dimension experiences a profound transformation of its own essence, exercising its ancient function and, yet, at the same time, assuming a novel anthropological character. The author supports his thesis by founding it, as he has always done before, on the works of a certain number of artists, whom he considers as the most meaningful of our age.

Keywords Technologies · Sublime · New media art · Externalisation · Connectionism

14.1 Introduction

A few years ago, the Contemporary Art Museum in Raleigh, North Carolina, organised an exhibition of five artists, introducing it as follows: "According to Mario Costa, the new technologies are creating conditions for a new kind of sublime: the 'technological sublime'. Data visualization in new media art transcribes raw data into a form of experience. It is driven by data, and its form follows from the optimization of efficiency in inputting data into a human brain. It is pragmatic, rational and logical. On the other hand, it is beautiful. It evokes emotions through its aesthetics.

M. Costa (✉)
Università degli Studi di Salerno, Fisciano, Italy
e-mail: costa@unisa.it

S. Chiodo, V. Schiaffonati (eds.), *Italian Philosophy of Technology*, Philosophy of Engineering and Technology 35,
https://doi.org/10.1007/978-3-030-54522-2_14

It communicates, distils complex ideas into simplicity and is thought provoking. In its consumption, it can be simply in delight and pleasurable".[1]

For the sake of brevity, I shall describe only three among these artists' works.

David Bowen's *Tele-Present Water* (2011) consists in an installation made by a flexible metal net that in real time gets information from the movements of seawater in a remote location, and translates them by waving just like the waves from which the information comes at that moment. It should be added that the transmitted data come from an adrift buoy that emits them while preventing the localisation of its position. Thus, "The water intensity and frequency is scaled and transferred to the mechanical grid structure, resulting in a simulation of the physical effects caused by the movement of water from this distant location".[2]

James Leng's *Point Cloud* (2012), differently from the weather forecast given in an abstract and static form, follows the changes in weather as they happen in real time, and makes them visible through a thin wire mesh that, suspended in the air like a cloud made up by points, changes according to the data it receives from the outside world. Thus, "*Point Cloud* is a sculptural form defined by a thin wire mesh, driven asynchronously by 8 individual servos controlled via Arduino. As whiteness of the hanging structure begins to disappear into the background, the viewer is treated to a constantly morphing swarm of black points dancing through mid-air. *Point Cloud* performs the data, dynamically shifting between stability and turbulence, expansion and contraction".[3]

In Ryūichi Sakamoto's *Forest Symphony* (2013–14) the bioelectric activity of a certain number of Japanese forests' trees, together with their photosynthesis, is monitored, collected by sensors, transferred in real time into the location of the installation, and transformed into sounds. As Sakamoto puts it, "Trees convert sunlight to energy through photosynthesis, meaning they are a genius at capturing electromagnetic waves. I want to turn that nature's cycle into music".[4]

These works effectively constitute a realisation of the theories about the "technological sublime" I have been articulating for the past thirty years, and represent an advanced end point in the process of aestheticization of technologies, which, as we shall see, started with the historical avant-garde movements.

In my opinion, in the last two centuries, technology has caused a corrosion of the very essence of art, which has become obsolete as a spiritual category, whilst at the same time creating the conditions of overcoming the arts towards what I have called "technological sublime".[5] In other words, as I have been arguing for many years, I

---

[1] See https://oss.adm.ntu.edu.sg/xqtoh1/wp-content/uploads/sites/34/2014/09/The-Sublime-Experience-in-Ecstatic-Truth_Data-Visualization-as-Art.pdf (last access on February 2, 2019).

[2] See http://www.dwbowen.com/telepresentwater (last access on February 2, 2019).

[3] See http://www.jamesleng.net/#/pointcloud/ (last access on February 2, 2019).

[4] See https://special.ycam.jp/interlab/en/projects/forestsymphony.html (last access on February 2, 2019).

[5] I identified for the first time the rising of this category in the '80s, and I gave a first definition of it in a symposium in Paris in January 1986 (see Costa 1986). *Technological Sublime* became, then, the title of a book (see Costa 1990), translated into French and Brazilian Portuguese, considerably

think that the *artistic*, as well as the other aesthetic categories historically related to it, such as beauty, expression, meaning, feeling that becomes image, symbol, genius, style, and so forth, are being substituted by a new age connected to that other important aesthetic notion, the *sublime*, as an effect of the present domination of technoscience. Thus, new technologies are, in my opinion, a new excess, and therefore make possible the manifestation of a new form of sublime that, after having been "rhetorical" in ancient times, "natural" in the eighteenth century, and "metropolitan-industrial" in modern times and until the beginning of our age, now becomes "technological". In other words, the "technological sublime" is a new age of the *aesthetic*, the only one that is possible after the sunset, now clear, of art, a new age characterised by the subject's extreme weakness and the impracticability of the realm of his/her meanings.

In this brief article, I shall try to show how all of this happened, and to analyse the conditions that made it possible.

Technologies have operated in different ways, and with profound effects.

14.2 Anthropology and Technology

Technologies have primarily operated on the anthropological level by causing the total externalisation of the human being, i.e., by everting her/him and making her/him live outside of herself/himself. This has happened in three fundamental moments, and in each of these moments change was caused by well determined and historically dominating media.

In the second half of the nineteenth century, *photography*, *gramophone*, and *telephone* started this process of human externalisation, which is still ongoing, and which has dissolved the internal human being and has determined her/his living outside of herself/himself. With these media, her/his image, voice, and *hic et nunc* have been technologically taken away, therefore her/his internal sense and roots in the world have been dismissed. Resistance to these technologies was very strong; it was perceived that they threatened the very essence of the human being, her/his more profound being. Technologies took away what characterised human beings the most, their body's ways of being: the image of the self as inseparably connected to the body, an image that only the mirror can reflect for an instant, the voice that only the body can bring around the world, their own *hic et nunc*, which is unique and non-transferable from instant to instant. Thus, resistance to these technologies was powerfully carried out: they were neglected, rejected, laughed at, considered as corrupting, immoral, and responsible for any atrocity, harmful to health, a constant reminder of our "you must die"; they subverted and deleted any trace of presence, deprived us of our spirit by disclosing our animality, dissolved our identity, and so

revised in a later edition (see Costa 1998). Finally, an article of mine, published in an American journal, was essentially focused on the *technological sublime* (see Costa 1991).

forth. But they were there, obstinately, irremediably, and more and more inescapably. And that is why a reworking of the imaginary, as an actual form of grieving process, was necessary to absorb and digest these technologies by integrating them within the already settled anthropological configuration, but by subverting it (as I tried to argue in my latest book, Costa 2018).

The radio, too, when it appeared, caused many forms of resistance. In the US the radio was feared to bring about the end of democracy, and Mussolini's and Hitler's use of it seemed to confirm this. The radio elicited the response of the masses, as suggested by Orson Welles's CBS programme in 1938.[6]

But with *television* things change, and it is this new medium that actually causes a newer and deeper form of externalisation. With television, it is not anymore just a matter of *fake news*, since the visual and acoustic real time highlights the "Welles's effect" and makes it conscious of itself. Now, people externalise themselves through communication and its main medium, i.e., *television*. Its *media-events* essentially mean a dissolution of the *self* and of its *identity* within a communicative collective consciousness that goes beyond any possible effect of the radio, and for the first time unequivocally manifests itself. Here people mindlessly abandon themselves to what Nietzsche calls "herd instinct": "for the longest period of humanity's existence there was nothing more frightful than feeling alone [...], one formerly experienced egoism as a painful thing, [...] the more the herd instinct and not the sense of self spoke through the action" (Nietzsche 2001, p. 115). "My idea is clearly that consciousness actually belongs not to man's existence as an individual but rather to the community- and herd-aspects of his nature [...]. Consciousness is really just a net connecting one person with another" (Nietzsche 2001, p. 213, 212).[7] Thus, television is not the radio, and works very differently: the radio shapes the "herd", television shapes the "abandoning of oneself to the herd", the "herd's pleasure".[8] Here what is at stake is really a collective consciousness, technologically shaped, which becomes protagonist by dissolving the *self* and making it resound in the same way in any person. Thus, here the "living outside of oneself" becomes specified as a "living inside the whole of the total communication".

[6] *The War of the Worlds* was a radio drama inspired by a H.G. Wells's novel, interpreted and directed by Orson Welles, and broadcasted by CBS on October 30, 1938, which made people believe that Martians were invading Earth, causing panic in an area of the US. Today, the critical consensus is to considerably resize the number of people that actually believed that what was being broadcasted was true. Yet, the programme proved that the radio had the power to impress a large number of people. This has been called "Welles's effect", and had actually been predated by a French radio programme by Pierre Cusy and Gabriel Germinet, *Mare-Moto*, broadcasted on October 19, 1924, which had caused an analogous effect to that caused by Welles's radio drama, even if reduced.

[7] It would be totally wrong to consider this *net* as "language", since Nietzsche wants to talk about a kind of ectoplasmic collective consciousness that pre-exists and founds language itself.

[8] These expressions are mine. In June 1981, 21 millions of people followed, at the same time and in real time, the story of Alfredino, who fell down an artesian well in Vermicino, Italy; and, in August 1997, television sent, so to speak, 3 millions of people from all over the world, all equally touched, to bring flowers to Lady Diana in London.

But the scenario changes again, and here we are in our age. Now, the dominating media are the *Internet* and mobile *phones*, and *communication* is substituted by *connectionism*.

These two media show up almost simultaneously, and hardly by chance. While the collective consciousness, as a communicative ectoplasm, becomes a body and objectifies itself into the Internet, mobile phones transform human beings into indistinct senders of signals without messages. The Internet is a real and impersonal hyper-subject that takes on itself the symbolic and its functions. It is able to contain it all, to produce it, and to indefinitely self-grow.[9] To be connected to the Internet means to get dissolved into an exorbitant symbolic, which is not ours anymore. Thus, while the Internet deprives the species of the symbolic by taking it all on itself, mobile phones externalise the human into a compulsive communication without a message,[10] because what really matters is connection, and not communication: this permanent injunction to communicate transforms the old principle of the "Palo Alto School", according to which *it is impossible not to communicate*, into a new principle, according to which *it is impossible not to be connected*, i.e., *nothing is possible but to get out of oneself*. Thus, connection compensates for the disappearance of the symbolic and the "inner listening", and substitutes them.

And the externalisation proceeds towards a goal that remains obscure and unpredictable: nets, interfaces, robotics and tele-robotics, total digitalisation, bioengineering and nanotechnologies, besides their possible and unpredictable combinations, and so forth, are making a world where the whole human being lives, in some ways dramatically, inside technology, and is becoming indistinguishable and radically conditioned by it.[11]

14.3 Art Inside Technology

It is necessary to start from a brief summary of the history of art from the avant-garde until today.

---

[9] I am thinking not only about what is already happening, but also about the big data and the Internet of Things related to the web, and, moreover, about when the Internet will host and connect the AI's work.

[10] And in the dictionary there are now new words: a "smombie", a term deriving from the fusion of "smartphone" and "zombie", is someone who walks on the street as a "zombie", looking at a mobile phone and sometimes wearing earphones. For these people, exponentially increasing, the Chinese city Xi'an has planned and built dedicated lanes, while around some Italian streets road signs indicating a proper ban have started appearing. Instead, "nerds" and "geeks" are solitary people obsessed with technology who more or less abandon social relationships in order to spend their time always connected.

[11] The people who die because of "selfies" are extreme cases, but extreme cases are precisely what help us understand (see Costa 2015). The very pulsion to communicate, discussed above and caused by the technologies at hand, is what turns all of us into empty transmitting points who receive signals.

After the Second World War, art's fundamental issue was still its own survival. The first avant-garde had severely destroyed it: Duchamp exhibited a urinal; the Dadaists shouted everywhere that art should be over; as for the Futurists, besides the will of destroying museums, libraries, and academies of all kinds, it was not clear what they wanted to do; painting lost its way; in Russia, *LEF* and *Proletkult* stated the sheer end of art; and, in Paris, Isidore Isou and the Lettrists claimed that all Western culture was obsolete and collapsing.

Indeed, art was still in danger of dying; yet, art, just like Misery in Robert Reiner's terrible film, should not die. Thus, in short, people made use of a strategy that continues to this day and that is increasingly effective, i.e., the strategy of identifying art with its own exhaustion and, ultimately, its own end. But first things first.

It was necessary to look the other way and start again precisely from painting. After the Impressionists, painting proceeded in the direction Maurice Denis pointed at in 1890. Indeed, Denis opened his *Théories 1890-1910*, published in 1913, with the well-known statement, or "first principle", according to which it should be remembered that a painting, before being a horse in a battle, a naked woman, or whatever representation, is essentially a flat surface covered by colours put together in a certain way (Denis 1920).

And so it was. From Gauguin to abstract art, painting was nothing but a surface covered by colours put together in a certain way, decorative, *fauve*, lyric, or variously geometric. Restarting from painting meant restarting from where painting had arrived.

Rosenberg's action painting (1952) and Greenberg's colour field painting, i.e., the two types of abstract expressionism, restarted from that point: Greenberg's flatness was Denis' flat surface without a horse in a battle or a naked woman.

Indeed, the new aestheticization of the surface could not be obtained at all from figuration, otherwise the avant-garde would have to be rejected. It was even harder to talk about beauty, which was impossible to revive after the avant-garde. What to do, then? Yet, even after having dismissed "beauty", "attraction" still remained. "Attraction" is what materials and signs immediately have thanks to their own constitution, their own immediate action on the nervous system, and the emotional resonance they cause: the colour grey is not a bright green, the zigzag line is not a circle.

As for colour, Hermann Rorschach's test is too well-known to be recalled. It is sufficient to say that the emotional reactions to being exposed to the colours of the ten tables constitute the most important element in analysing the psyche's deep levels. Back in the 40s, the test was considerably widespread in the US, and its existence was well-known also beyond experts. Rorschach died in 1922 at 38, one year after having published his book, and, in his test, colour was only one of the elements to analyse.

In 1949, another Swiss psychiatrist, Max Lüscher, developed another test, in which the analysis of personality was entirely founded on colour: his "colour test", which analytically described the concussive effects of any single colour, immediately circulated worldwide.

Signs' action on the nervous system, due to their immediate physical configuration, is also one of the most persuasive points of Susanne Langer's aesthetics: if I draw a line in a certain way on a surface, then the sensations and virtual solicitations I get from it are different from those I get from a different line, since lines "express vitality in what they themselves seem to 'do'" (Langer 1953, p. 63).

Thus, in order to let painting survive, abstract expressionism's strategy, as well as that of its articulations, was to foreground the primordial and physiological level of colours and lines of force and to delete representation, still obtaining an effect somewhat aesthetic.

In Kantian terms, we may say that abstract expressionism substituted *beauty* with *attraction*, i.e., used neurosensory solicitations coming from physical elements, such as colour and lines, in order to establish a pleasure purely and only perceptive (of course, virtually perceptive), which not only has nothing to do with beauty, but also (and more precisely) with art.

This use of the "matter of satisfaction"[12] (Kant 2000, p. 5, 223) as art, or rather as the most authentic art, is, a little before abstract expressionism, a fundamental principle of Jean Dubuffet's work and theories. When, in 1968, Dubuffet published the book *Asphyxiante Culture*, it was not a matter of following the trend. Rather, his intolerance for the global organisation of society, typical of the spirit of '68, had already manifested many years before, just after the Second World War, when, because of this intolerance, he had started collecting "artistic" objects that had nothing to do with Western art: he identified an artistic radical alterity in artefacts, sometime even very badly made, created by prisoners, lunatics, street artists, children, and primitive people. *Asphyxiante Culture* is the *summa* of his anticultural and antisocial beliefs, according to which culture, as opposed to what is usually believed to be, is restrictive, limiting, and obscurantist, since it lacks the pleasure for anonymous innumerable germination (Dubuffet 1986, p. 15). Thus, if culture and art are all of this, then it is necessary to look elsewhere, and, believing that barbarity is a value to preserve (Dubuffet 1986, p. 56), to search for an *art brut*, generally not contaminated by culture, where it is and hides. This use of the "matter of satisfaction" to create art, which is the essence of Dubuffet's *art brut*, actually coincides with the strategy used by painting in order to survive, from 1945 to the early 50s, throughout the world, from Europe to the US.

The principle is the same, and includes both works in which figuration is still present in a way (see the CO.BR.A Group, Dubuffet, or Fautrier, as well as some American Abstract Expressionists such as de Kooning) and pure abstraction (see Franz Kline, Motherwell, and, differently, Rothko and Barnett Newman).

Yet, the thing is that, once the "matter of satisfaction", i.e., colour and signs, are established as art's autonomous principle, not only any essential difference between both the above-mentioned artists and the above-mentioned artistic movements dis-

---

[12] "Taste is always still barbaric when it needs the addition of charms and emotions for satisfaction, let alone if it makes these into the standard for its approval. And yet charms are not only often included with beauty [...] but are even passed off as beauties in themselves, hence the matter of satisfaction is passed off for the form" (Kant 2000, p. 5, 223).

appears, but it is also impossible to distinguish what art is from what art is not. The above-mentioned artists, together with many others, such as alienated people, children, and primitive people, use the "matter of satisfaction", and use it in the same way, by working on a purely neuro-emotional level, and the works of the former or of the latter have no inner reason that allows to distinguish what art is from what art is not.

For this reason, then, painting ran into a dead end.

But art, just like Reiner's Misery, should not die. Thus, if it was the painting's medium that disappeared, and could not be used anymore, then it was necessary to get out of the medium, get out of the surface, neglect Rosenberg, Greenberg, and the whole abstract expressionism, get out of painting, and turn it into something else.

Robert Rauschenberg is the author of this major breakthrough concerning the removal of the surface, as well as of painting, which is a real turning point within the history of contemporary art. This tendency to abandon the surface is extremely clear in Rauschenberg, and it is represented by the progression, in the 50s, from the *White paintings* to the *Combines*. In dozens of *Combines*, Rauschenberg doubles the surface in the space and puts objects that exceed the paintings, dismissing in this way both painting and the history of painting at the same time: the painting's colours continue on a chair leaning against it (a chair that in this way becomes a part of the painting itself), two umbrellas exceed the painting and open up in the room; in another painting the umbrellas are substituted by two fans, the centre of the painting is constituted by a stuffed eagle, as well as a ladder, washing lines, a bucket, various wheels, birds and little birds, and so forth, all constitute something to look at. Indeed, with Rauschenberg the history of painting as fine art ends, and the history of art as such starts, as Rauschenberg would say: "All this is art and art only, even if I still use painting". But it is precisely this "still using painting", however it may be used, that still includes Rauschenberg within the realm of contemporary art. His *Combines* still have a sensory impact, still have to do with the *aisthesis*, with an aesthetic kind of perception. But it is for the last time.

After having exhibited, in 1962, the *Campbell's Soup* silkscreens, which had already been quite sensational, Andy Warhol exhibited, in 1964, the *Brillo Box* at the Stable Gallery in Manhattan. *Brillo Box* consists in a stack of wooden boxes empty and perfectly identical to the original ones, which obviously were full and made of cardboard. This difference is not visible, and, actually, Warhol's boxes and the ones you can find in an American supermarket are definitely indiscernible. But his boxes were expressly made for the exhibition, and partly different from the original ones. Thus, they could still have something artistic. To avoid any misunderstandings, Warhol, in the pop art great exhibition in New York, in 1964, signed a whole stack of original *Campbell's Soup* cans, which was meant to say: "As a matter of fact, I want to give a meaning precisely to the supermarket boxes, and, by doing so, turn them into art". Indeed, the audience wonders why Warhol did what he did, and what he wanted to say; that is, what is the *meaning* of what one sees, and why it is art. This was the obvious discovery made by Arthur Danto, the great American art critic of those years.

There is a consensus, even influential, that contemporary art does not make any sense: Jean Baudrillard, to name but one, in the famous article published in *Libération* on May 20, 1996, denounced contemporary art's meaninglessness, its nonsense.

Actually, now the exact opposite is true: contemporary art can survive only by giving a meaning to what it does. Between *aisthesis* and *meaning* there is an inverse correlation: the more the *aisthesis* disappears, the more the *meaning* is necessary and increases, and, in certain forms of "conceptual art", this happens all the way up to the total removal of the *aisthesis* and the exhibition of pure and simple *meaning* only.

With the *Brillo Box*, Warhol performs something that, if considered from within the realm of art, puts an end, at the same time, to art and the history of art, since it changes its very concept. Warhol's performance was to be considered as extraneous to art, and art should have gone its own way. But art was already at a dead end, and Warhol's could have been a solution, to be seen as a metamorphosis inside of the very process of art's evolution.

Thus, if art is not *aisthesis*, but *meaning*, then anything can be meaningful, meaning can be obtained in any way and through whatever means, anybody can do this, and art is reduced to an attribution of a meaning: making art means giving a meaning to whatever. But, then, if the new essence of art is *meaning*, and if everybody can give a meaning to whatever, then how is it possible to identify the artist? And, then, if the *aisthesis* disappears, and if anybody can put in place a meaning, then the great artist should rely on something else, and this thing cannot be anything but the *information* with which the meaning is communicated: indeed, it is a matter of turning *meaning* into *information*.

Now, from a formal and functional point of view, only an excessive, unexpected, irritating, obscene, nauseating, blasphemous, and so forth, *meaning* can make meaning itself informatively effective. But the *meaning/information*, even if correctly made in this way, remains inefficient without a communication system that can impose it by circulating and disseminating it. And this is how we get to contemporary art: *a work of art is an excessive, unexpected, irritating, obscene, nauseating, blasphemous, and so forth, meaning/information that is delivered by a more or less powerful communication system, which can impose it.*[13]

[13] Someone drawing a spiral on the Great Salt Lake shores, or someone else packaging Little Bay cliff; someone hurting themselves with rose thorns, or someone else dirtying themselves with dismembered animals' blood; someone telling us that the chair in the picture is not the chair on which we sit down, or someone else making us enter a stable with horses and their excrements, or making us chat in an art gallery, and so forth, and getting worse and worse, up to sliced animals in formalin, or a frog put on the cross instead of Jesus Christ… All these things surely have a meaning, effectively delivered, but saying that they actually have anything to do with what art has been for some millennia is simply a gross begging the question. To the above list, I add a list of works of contemporary art given by Francesco Bonami, a well-known art critic who seems "regretful": "pots of boiled mussels, little cans of shit, empty shoeboxes, urinals, neon tubes, coloured iron boxes, big rusty iron slabs, coat racks, dirty plates with food trash, cut canvas, burned sacks, unusable loudspeakers, pieces of felt, pieces of rotten butter or some sausages, a dead hare, a safe with

As a matter of fact, an artist will be more or less great according to the quantity of *information* that her/his work contains and to the power of the *communication system* that supports it.[14]

What is described here is nothing but the drift reached by what we knew as art, a drift caused by all the technologies that were developing.[15]

And this is to say that art has no meaning anymore. Consider, for instance, the *décalage*, i.e., the difference existing between the architecture of contemporary art museums, actually projecting towards the twentyfirst century, and the rubbish constituted by everything inside of them, i.e., the so-called contemporary art. Indeed, art can survive as a *bricolage* of hundreds of thousand artists, who are more than the sand in the sea,[16] or as a billionaire "nothing machine", which nobody can ever remove, but which has irremediably lost any interest and any active meaning.[17]

---

holes, etc." (Bonami 2017, p. 100, our translation). I also add the list I provide in Costa (2010). And for another list, which is more truculent and revolting, I refer to Strassoldo (2010). Yet, I specify that both Bonami and Strassoldo, in their lists, unforgivably confuse and mix up avant-garde movements and contemporary art, which does not work at all and is totally wrong.

[14] On October 22, 1995, *The Independent*, a newspaper based in London, published an article, *Modern Art Was CIA "weapon"*, by the journalist Frances Stonor Saunders, where she basically said that the CIA was behind the success of the American "abstract expressionism" and of the exhibitions that circulated it all over the world. A CIA former agent, Donald Jameson, broke the silence and told everything: after 1947, culture was to be a weapon in the "cold war" against the Soviets; the Americans were to oppose the rigid control of culture, exercised by the Russians, through the absolute freedom of their intellectuals and artists. The operation concerned all culture, but, as for art, the strategy would have worked at its best: the group of Abstract Expressionists was made up by anti-government, anarchic, and pro-communist people, and leading them to success against the tastes of the Americans' majority, Truman included, and at a time when "McCarthyism" was popular would have proved that freedom was really the greater good for American democracy. Thus, the CIA identified a group of artists that could work to establish a great American art, built Pollock's myth, brought to the MoMA a lot of money, and entrusted it with the promotion and the international exportation of the "abstract expressionism" through any means, especially through great exhibitions in Europe. After her 1995 article, Stonor Saunders published a big book in 1999 (see Stonor Saunders 1999), which extended what she stated about art to all the other cultural areas, but, basically, except for the extreme and impressively detailed information, often embarrassing also for our intellectuals, did not add any new element to the 1995 article. Since then, the system has kept working by abandoning political goals, and becoming a pure and simple luxury market where especially art galleries' multinational chains, contemporary art museums, and more or less stupid or money-maker collectors operate.

[15] I tried to show the role played by technology in this process in Costa (1990); Costa (1999a). I apologise for often referring to my books, but, since some of my remarks may seem groundless and unjustified, I should necessarily refer to the publications where these remarks are argued at best.

[16] Which at least is useful in order to reduce unemployment.

[17] Which now actually seems an open secret, in that it is declared in films (*The Square* 2017, which follows other older films such as *El artista* 2008), and in that even a great critic and international curator writes a book that "wants to tell about the end of contemporary art" (Bonami 2017, p. 4, our translation), and ends by saying that "contemporary art is over. Go in peace" (Bonami 201, p. 132, our translation).

14.4 The Technological Sublime

The works previously introduced as included in my concept of *technological sublime* are actually the end point of a century-long process.

The start dates of this process may be 1920, with Gabo and Pevsner's *Realistic Manifesto*,[18] and 1923, when Itten was substituted by Moholy-Nagy in the "preliminary course" at the Bauhaus.[19] Since then, and with these figures, starts something that joins art, on the one hand, and science and technology, on the other hand, and that does not have anything to do with art anymore. It is not just a matter of art taking into consideration what goes on in the realm of science, as it happens with futurism or cubism, where painting remains what it used to be, since only the image of the surface changes, whereas the very artistic object remains what it used to be. Instead, with Gabo and Moholy, the object's very nature changes, i.e., the artistic object becomes something else: Gabo and Pevsner's sculptures or Moholy's *Light-Space Modulator* break up with the whole tradition of art, because they are art, science, and technology joined together.

After the Bauhaus, and forced by Nazism, Moholy, in the 30s, emigrates to the US, where he founds the New Bauhaus. Then, together with his pupil and friend György Kepes, who follows him to the US, turns it into the Chicago Institute of Design; and Kepes, since 1967, founds and directs the Center for Advanced Visual Studies, which will become a reference point at the MIT for generations of artists from all over the world, who variously work on joining art, science, and technology.

In the 50s and 60s, there was a considerable intrusion of science and technology into art, with certain forms of *kinetic art* and *programmed art*, but only the diffusion of video and computer caused a wide and manifold "technological art": video art, computer art, electronic music, computer poetry, and so forth, interested a large number of artists from all over the world, and this continues to be the case today.

---

[18] In August 1920, the *Realistic Manifesto* appeared on the walls of the streets in Moscow, written by Gabo and signed also by his brother, Anton Pevsner, where "realistic" did not mean any figurative adherence to reality, but the fact that art would have realistically and scientifically investigated the "kinetic rhythms" and energies inside the matter: "The line plumb in hand […] the mind rigid as a compass, we are building our works as the universe builds". And Herbert Read confirms that Gabo and Pevsner's particular vision of reality comes from the structure of the physical universe as it is revealed by modern science, therefore the artist's intellectual vision comes from modern physics (see Read 1918). What matters is not that art is inspired by science, which in one way or another has always happened, but that this time inspiration essentially changed the artistic *object*, which became something else.

[19] Gropius decided to attract to the school the most decisive strengths of constructivism. In 1923, he appointed Moholy-Nagy to teach the preliminary course (see Moholy-Nagy 1975). When Moholy-Nagy entered the Bauhaus, it was dominated by "mystical" tendencies: Kandinsky, Klee, Itten. In 1923, Gropius, founder of the Bauhaus, understood that the school needed a radical turn: he expelled Itten and assigned Moholy-Nagy the preliminary course, which at the Bauhaus was a fundamental stage of the education programme. Moholy's wife, Sibyl, noticed the contrast between before and after, and reports that Klee was seen as someone who used to make spiritual samovar and intellectual handles (see Moholy-Nagy 1975).

Yet, a basic misunderstanding was involved in all this. People talked about the "humanisation of technologies", "new artistic tools", "new languages", and so forth, without realising that the very essence of art had changed. "Technological artists", or at least a large number of them, believed that nothing had changed, and that they could keep doing through the new media the same works they used to do through the old ones, therefore they put inside computers and electronics their narratives and feelings. For instance, with the medium of video, it was not clear that mimesis or narration are not its specific and essential qualities. But that was not always the case. Thus, depending on the level of "artistry" meant to be inside video, there were "artists" such as Robert Cahen or Bill Viola, on the one hand, and such as Stephen Beck or Steina and Woody Vasulka, on the other hand.

However, all of them, aware of it or not, did nothing but allowing technologies to be their masters. Indeed, only someone understood it: "We artists, working through these media, are not here because machines have been created, but because machines need us in order to live [...]. I am curious to know what all this means [...]. I can see that, in the long run, technology will become very complex and rich in challenges. Yet, I do not know whether it will need us".[20]

But let us go back to the original works. They put together all the current resources of information and telecommunications technologies, being a kind of synthesis of *aesthetics of communication*[21] and *technological sublime*, and make visible something that definitely breaks up with art and introduces a new anthropology.

But what are these works really making visible? At a first level, they turn into images the computer data received in real time, but this is not sufficient. The electroencephalogram makes visible the brain's electrical activity, its operations as living matter, as the "data visualisation", which operates in this case, shows the mental operations that activate it, makes visible the mind's basic symbolic activity, its logic, the abstract intellect's operating: indeed, mathematical and logical procedures become visible, through an unconceivable breakthrough passing from the level of categories to that of perception, and by subverting the topology and gnoseological dynamics of intuitions, schemes, and concepts.

Here, the relationship between thought and perception gets subverted: now, thought has to do with perceptive data that are not the real data, for they refer to other data that make them possible; but these second data are not immediately perceived, are not perceptible, and can only be thought of, in that the perceptive data, i.e., the real perceptive data, exist only as thoughts. Thus, thought is in conflict with itself.

---

[20] Woody Vasulka in Peter Kirby's *Binary Lives. Steina and Woody Vasulka*, video, 1996.

[21] The *aesthetics of communication*, a few years before the *technological sublime*, expressed the necessity of identifying and putting in place ways of aestheticization of remote simultaneity, as well as of the new experiences of space-time made possible by it. I stated it for the first time in October 1983, in a "manifesto" signed at the same time by the French artist Fred Forest and by the Argentinian artist Horacio Zabala. After several other publications, in Italy and abroad, a dossier was dedicated to it in an issue of a French journal (see Art/Press). Afterwards, together with others, I reconstructed its history in Costa (1999b).

As a matter of fact, it is a kind of revenge on the part of technology, since technology, previously considered as something outside of philosophy, now is put at the very centre of philosophy itself, making it rethink fundamental ontological issues (see Costa 2012, 2015; Benso 2017).

But, here, my aim is talking mostly about art, therefore let us keep talking about it.

These works dissolve all the notions on which art has been founded: they are not metaphors, languages, meanings, symbols; they do not make reference to anything but themselves and what they do. They create a new state of affairs, renew the way of being in the world, simply put us in another situation, in another being there, which was not at our disposal previously, but which is now made possible by the current information and communication technologies. These are able to put the human species in an absolutely new condition, i.e., in a situation in which here and there, this place and that place, being there and not being there, movement and stillness, and so forth, become all present at the same time, blended in a new form of experience made of perception and reason, which continuously make reference to one another by remaining blended. Artists make this new situation perceptible and give it visibility, therefore this situation assumes an aesthetic dimension that has nothing to do with *art* anymore and puts in place a new form of *sublime*.

If I see a metal net moving just like the sea waves are moving somewhere else, if the movements of a tropical plant's sap are turned for me into sounds I can hear, if somebody can make the plants communicate, and make this communication perceptible to me, then all this—which is not art anymore, which has no form, because its form dissolves into space, and which puts in place technology's unconceivable power—causes to me an indefinable apprehension, constituted by perception, reason, and emotion blended together, which I identify as *sublime*, maybe that sublime characterised by calm admiration Kant used to call *noble sublime*.[22]

Here, there is no form to contemplate, this is not a work meant to mean something, as it just wants to *be* something, to put in place a *way of being* previously non-existent in this world, since human nature could not perceive it, which is now technologically possible instead.

By now, our world is a mystery become obvious and not perceived anymore. The mysterious character of electricity, to give one example, was continuously perceived and expressed by Proust, who, at the same time, used to stigmatise its trivialisation by making reference to the telephone as a supernatural instrument before whose miracles we used to stand amazed, and which we now employ without giving it a thought; and this happens because habit quickly deprives the sacred forces we are dealing with of their mystery (see Proust 2003). The "artist"'s role, once she/he became an inquirer of the profound structures of our current being in the world, is that of confronting us with all that is technologically possible, by making it manifest as something we can perceive, and is that of making us feel the sacred forces that

[22] "The sublime is in turn of different sorts. The feeling of it is sometimes accompanied with [...] quiet admiration [...]. I will call [...] the second the noble" (Kant 2011, p. 2, 209).

now mysteriously permeate our world, as well as making us dive in the ecstasy or the terror of the *technological sublime*.

English translation by Simona Chiodo, revised by Diletta De Cristofaro

References

Benso, S. (Ed.). (2017). *Viva voce. Conversations with italian philosophers*. New York: Suny Press.

Bonami, F. (2017). *L'arte nel cesso*. Milano: Mondadori.

Costa, M. (1986). *Le "sublime" et l'estétique de la communication*. In *Ecole Nationale Supérieure des Beaux-Arts* (pp. 11–14). Paris: Artcom.

Costa, M. (1990). *Il sublime tecnologico*. Salerno: Edisud.

Costa, M. (1991). *Technology, artistic production and "aesthetics of communication"*. *Leonardo, 24*(2), 123–125.

Costa, M. (1998). *Il sublime tecnologico. Piccolo trattato di estetica della tecnologia*. Roma: Castelvecchi.

Costa, M. (1999a). *L'estetica dei media. Avanguardie e tecnologia (1990)*. Roma: Castelvecchi.

Costa, M. (1999b). *L'estetica della comunicazione. Come il medium ha polverizzato il messaggio. Sull'uso estetico della simultaneità a distanza*. Roma: Castelvecchi.

Costa, M. (2010). *Arte contemporanea ed estetica del flusso*. Vercelli: Edizioni Mercurio.

Costa, M. (2012). *Ontologia dei media*. Milano: Postmedia.

Costa, M. (2015). *Dopo la tecnica*. Napoli: Liguori.

Costa, M. (2018). *L'uomo fuori di sé. Alle origini della esternalizzazione. La fotografia, il fonografo e il telefono nella Parigi del XIX secolo*. Milano: Mimesis.

Denis, M. (1920). *Théories 1890-1910*. Paris: L. Rouart et J. Watelin.

Dubuffet, J. (1986). *Asphyxiante culture (1968)*. Paris: Les éditions de minuit.

Kant, I. (2000). In P. Guyer (Ed.), *Critique of the Power of Judgment (1790)*. Cambridge: Cambridge University Press.

Kant, I. (2011). In P. Frierson & P. Guyer (Eds.), *Observations on the feeling of the beautiful and sublime and other writings (1764)*. Cambridge: Cambridge University Press.

Langer, S. (1953). *Feeling and form. A theory of art*. New York: Charles Scrinber's Sons.

Moholy-Nagy, S. (1975). *Moholy-Nagy. La sperimentazione totale (1950)*. Milano: Longanesi.

Nietzsche, F. (2001). In B. Williams (Ed.), *The gay science*. Cambridge: Cambridge University Press.

Proust, M. (2003). In C. Prendergast (Ed.), *Search of lost time*. London: Penguin.

Read, H. (1918). *Gabo/Pevsner* (p. 11). New York: Museum of Modern Art.

Stonor Saunders, F. (1999). *The cultural cold war*. New York: The New Press.

Strassoldo, R. (2010). *Da David a Saatchi. Trattato di sociologia dell'arte contemporanea*. Udine: Forum Editrice.

Chapter 15
Ontology of the Virtual

Roberto Diodato

Abstract The article develops an ontology of the virtual, here interpreted as a new entity which appears in the world thanks to new digital technology. Being an ontological hybrid, the virtual comprises body and image, object and event, internal and external, artificial and living. Furthermore, the virtual entity is the result of the interaction between computer programming and the human body augmented by technological prostheses and, since this entity only exist in a dimension of interactivity, it shows the priority of the category of relation.

Keywords Virtual · Interactivity · Relation · Digital · Ontology

15.1 Body/Image

By "virtual" we can understand in the first place and in a broad sense a body-image (Diodato 2013b, 2018): an interactive digital image, the becoming phenomenon of the binary representation of an algorithm in interaction with a user (for the other definitions of "virtual" see Grimshaw, ed., 2014, pp. 17–125). What I am talking about are all those digital objects/environments with which users can interact through the devices we can connect to computers, these sometimes even being bio-robotic prostheses which allow for high levels of immersiveness. Sometimes users interact with such digital environments by means of avatars, virtual alter-egos who appear to perform actions within these environments and to produce transformations within them; some other times the users' spectator function coincides with their being actors in the situation. Now, the whole "aesthetic" transformation or modification produced by users in digital or virtual environments are possible inasmuch as

R. Diodato (✉)
Università Cattolica del Sacro Cuore, Milan, Italy
e-mail: roberto.diodato@unicatt.it

S. Chiodo, V. Schiaffonati (eds.), *Italian Philosophy of Technology*, Philosophy of Engineering and Technology 35,
https://doi.org/10.1007/978-3-030-54522-2_15

the sense-based images (involving sight, hearing, touch, etc.) they perceive or produce are nothing but the several phenomenon versions of an algorithmic matrix, nothing but the several possible aesthetic actualisations the program allows (Diodato 2014). However, it is interesting to remark that the degree of interactivity of digital objects varies based on whether the interaction relies on "stiff" algorithms—which pre-order all interaction—or "flexible" ones—which "learn" as they go and change according to the interaction itself.

Endowed with a grammar, such an image-language implies a characteristic spectrality which affects the visible-invisible relationship and structures the modalities of the use. From this point of view, the digital body-image which can be multisensory is not simply image-of, not only mimesis of an, identifiable or not, thing or image. It therefore does not have a simulacral essence. On the other hand, it is not even an icon or an original image; it is rather a genetic relational form belonging to a multiple rendering system. The virtual entity is in fact a good example of system, of a complex system which is able to express what pertains to systemic ontology. Within such an ontology, "system" is by no means reducible to "element", since, if each "element" is a system, either "systems" are positively conceived or we have an infinite regress. Moreover, provided that the "element"—both taken simply as element or in its component function—is what is ontologically equal to the "object", "system" then does not mean "object", no matter how we understand the latter—be it the combination of material substratum and properties, substance and accidents, a bundle of properties or a bundle of tropes—even if we were to take it as a unity of elements *obviously interrelated*. Even in this latter case, in fact, relations are taken as properties of the elements-parts forming the systems, based on which what one may call "system-object" turns out to be endowed of *relational properties* ("According to systemic thinking, an object should be described as an organization of parts connected by relations, and having properties that parts do not have; such properties are called 'emergent', or 'second level', or 'systemic properties'", Urbani Ulivi 2018, p. 195). Differently, the system can be understood as a relational structure, and the relations can be assessed in themselves, with no need to transform relations into elements. "Relations" are here understood in contrast to "relational property", as the relation without which there is no identity, therefore no existence. What allows us to do that without hypostatising the relation, and without immediately taking it simply as causal relation, can be better grasped, possibly, in reference to a few entities which make best display of their systemic nature: the virtual entity makes, in this regard, as we shall see, for a remarkable case (Diodato 2010, 2013a, 2015, pp. 47–65).

Since it is made up of ordered sequences of binary units, or, in other words, of character strings developing on different levels of a syntax, the digital image is not, we could say, exactly "image", but rather body-image. Interestingly, the coincidence of these strings and their sensible appearances, at present mainly sound or visual, are generally perceptible. Now we know that discrete sequences can also render continuous undulatory phenomena. A thin body in a non-continuous world of data points can be conveyed as fluidity and density and even saturate the perception. Informationally or in the formal terms of programming languages, the virtual is for sure an electronic body (Hansen 2006), hence—metaphorically—an atomic aggregate, but the digitalisation process makes it peculiarly light: a compound of sometimes remarkable quantities of extraordinarily fast transmittable data—the highest speed permitted by the physical limits—it can assume various incarnations, both structurally identical and phenomenically different. As an hybrid entity, the virtual is essentially interactive. This is a delicate point, but at least it conveniently allows us to exclude from the notion of virtual body-image those simply photo-graphic or tele-vision digital images which entail a simply passive action, which does not affect the properties of their appearing and do not permit a retroactive interaction on the structure of the computer memory, in other words an impact on the matrix. The range and different degrees of interactive operativity are obviously very large, as well as the meaning of the notion of "virtual body-image".

Moreover, the notion should be distinguished from traditionally understood images. We can say, for instance, following Sartre's intuition (Sartre 1986, pp. 14–34), that any consciousness of an image is consciousness in its non-realising function, which has its noematic correlative in the imaginary. Therefore, being "image" is clearly ontologically distinguished from being "perception". Differently, in the virtual environment, it is not clear how "having an image" could be distinguished from "having a perception". It is equally not clear *what* can be called an image: the virtual entity, inasmuch as it is intentionally constituted, is structurally ambiguous. In a virtual environment, it is not true that what is conventionally called image is immediately given as such to reflection, but what happens is that perception and imagination rather mysteriously fade into one another, because the percept, faced as perceptual intention, is not properly external, it is neither *in* consciousness nor *in* the world; it is external-internal, it is itself an image, which is not a thing-image of the world (e.g. a painting, a photograph, a cinematic image, a digital image like TV images), but a body-image, constituting itself in the very interaction. My goal is to approach the strong notion of virtual entity from the point of view of the novelty of its ontological statute.

15.2 Metamorphosis

From a conceptual point of view, the problem of an interactive metamorphosis does not concern the possibility of registering perceptions in users—which rather depends on the quality of the sensors, as well as on a reductionist background hypothesis—and in principle it is not even a matter of management by the processor of a data bank of morphologies connectable with these mutations. What is instead at stake is the kind of relationship that can be established between the alterations of the user and the generation of video and audio forms. It is possible to plan, through an interface scheme, a punctual translation system able to cause a system of variation-substitution of the configuration-sculpture; while it is harder to have a system that can make the forms of the work evolve in relation to the interaction with the user, there is an evident watershed between the kind of interaction that takes place in the established limits of a materiality which is already given, in our case a data bank of even remarkable size, and an interaction that can affect—within the metaphor of the sculpture—the programmed matter, making it evolve in states which were not predictable.

In order to grasp the meaning of the expression "interactive metamorphosis" when it comes to virtual entity, it is worth digging deeper into the meaning of the term "metamorphosis". Based on ontological premises establishing that what exists, a being, in the observable realm, is the whole of matter and form together (*synolon*), under no circumstances what in such a compound one calls "matter", "potential aspect", or "principle of indeterminacy" can be mistaken for elements or parts of an observable being. Matter and form are metaphysical principles and what is commonly known as form has been given many names (e.g. *morphé* or inherent form, *eidos* or form geared toward, *energheia*, *entelechia*, etc.) according to the situations/functions that one sought to describe. Granted that beings are the whole of matter and form together, in the case of a being that is also a system, it would be a mistake to take form as the organisational principle and matter as the parts/elements of the system to be organised. It would be an even greater mistake to strangely reverse things and give matter the function of form, as to make the *hypokeimenon* act as a principle of permanence and the parts as a principle of variation. Matter is better envisaged as the source of potentiality, in other words, as *dynamis* pointing to the positive and real ability of the *hyle* to take but also to lose the *eidos*. This would ultimately account for the metamorphic tendency of beings and their overall dynamics in terms of *energheia* or form as actualising process of what the *synolon*, with respect to its potential, is still missing. As individuating process amounting to the achievement of all the possibilities of the *synolon*, the actualisation is continuous. Based on this philosophical perspective, matter and form are one and the same thing, the former potentially, the latter actually. Depending on which theoretical path one follows, one could take the *synolon*, what one calls being, as the whole of potentiality and actuality, or the whole of matter and form as meta-stable substrate to what is not yet but could be, or as indeterminacy which is progressively organised thus achieving possibility. The not yet formed is already a form, although still undetermined as to what form it could have. The limit of what is in the making is given by beings themselves, inasmuch as their metaphysical composition are also the boundaries determining its power. In other words, given that matter is potentiality, deprivation is non-potentiality, hence limit to power, and, in this respect, somehow form. It is therefore not about achieved possibility, but about boundaries, perspective, theoretical cuts, potencies and impotencies, pertaining to that distinctive indeterminacy, always only provisionally determined, that are beings. The adverb "provisionally" stands for the impossibility to mark down an instant of time *t*. The term "actuality" stands for a process and what disappears is always something that is absent, the determinate absence of what is potentially determined or not actually determined. The metaphysical principles of beings are then an unknowable undetermined (differently named according to different perspectives: *hyle*, *dynamis*, *hypokeimenon*, etc.), the entailed deprivation (*steresis*) rooted in it, and the actualising (*eidos*, *morphé*, *energheia*, *entelechia*, etc.) that it continually and variedly determines. What is actually determined is also potentially determined, which means that it can take upon itself a given form, deprive itself of that form and so on, in the seamless process one calls metamorphosis. Composition and dissolution: a twofold tendency intrinsic to the undetermined, what has no form, what is potentiality *dynamis*. Toward form and toward deprivation of form: incessant metamorphosis aiming at construction and destruction.

15.3 Mimesis

The generation of "virtual reality" means the generation of the possibility of experiencing an environment—identified as an "environment" by a set of "virtual bodies" which are not bodies belonging to the environment or in the environment, but bodies coinciding with it—which is able to produce perceptive experiences in the user. By "virtual-reality generator" one can then understand a machine which is able to make the user experience such an environment, in other words, to render the environment in a situation. This amounts to say that a virtual-reality generator *could be* conceived as a generator of possible sensory perceptions, and precisely as a generator of sensory perceptions (visual, auditive, tactile, olfactory, etc.) which are able to simulate an environment-situation that we could pre-philosophically and with sufficient accuracy define "real". In other words, a virtual-reality generator should imitate "perceptive faith" that seems to be presupposed in our daily dealing with the world. In this regard, the reader could be under the impression that this account relies on a restrictive and after all not very interesting definition of environment and, accordingly, of virtual body-image, one that is keen on equalising virtual reality and its reproduction, or even considering the virtual as a part of a mimetic project. An environment that is identified as virtual because of its capability to reproduce a real situation turns out to be accurate inasmuch as it is able to answer each possible action of the user in the desired way. Its accuracy, then, does not only depend on the experiences the users actually perform, but also on the ones they *could* perform. The evaluation of the "sufficiency" of this accuracy is clearly a problem: can reality be reproducible to the last detail? Can a "perfect illusion" be built? Supposing the user is able to make free choices, in the sense of freedom of indifference, reproduction is impossible, because such choices are not computable. If we stick to another metaphysical hypothesis, supposing, for simplicity, that choices are the result of an endless causal series—where the idea of series itself is reductive and inadequate—the reproduction of reality will be all the more efficient the more the elaborator is able to calculate the possible actions and reactions of the user, and consequently to prebuild potential interactions among the virtual environment-bodies. So the more such computation tends to infinity, and accordingly such algorithm is phenomenalisable, the more virtuality reproduce reality. From this point of view, the virtual environment is an imperfect Spinozian machine, a set of relations that constitutes a prospective coincidence of freedom and necessity. Such coincidence could be possible only in an essentially incomputable, endless, causal network. As a consequence, while the virtual environment aims to produce the experience of a pervasive and persuasive immersion, such an experience is also aware of its own particular ontological statute. It indeed appears as a mimetic prospective, and not as a perfect mimesis, and in my opinion it is actually this limit, void, lack, that opens the artistically relevant possibilities of the virtualisation of imagination (Grau 2003; Lopez 2010).

15.4 Internal/External

The topic of the instant of time clearly comprises that of the countability of the continuum, hence the topic—essential to our question—of the relationship between number and reality or, in other words, between programming languages and electronic matter. Furthermore, if a virtual body is and is not itself in place and time, since its eventualisation depends on the interaction with the user, can we now claim that reality is interactive *in the same way*? In reaction to these questions, David Deutsch writes: "What may not be so obvious is that our 'direct' experience of the world through our senses is virtual reality too. For our external experience is never direct; nor do we even experience the signals in our nerves directly—we would not know what to make of the streams of electrical crackles that they carry. What we experience directly is a virtual-reality rendering, conveniently generated for us by our unconscious minds from sensory data plus complex inborn and acquired theories (i.e. programs) about how to interpret them" (Deutsch 1997, p. 120). What Deutsch states is nothing but another form of transcendentalisation of the empiric: what we are psychophysically affects our "direct" experience of reality, meaning by this the constitution of a sense full environment. Such an outlined scenario also requires the clear formulation of a theory of knowledge and a corresponding ontology. It will suffice here to say that the virtual body does *seem* to have at least one quality which is different from the ones of the bodies we usually call—on a common sense level and with the languages of theories—real. One could say, for instance, that reality is not interactive in the same way as virtual reality is, and that *real* bodies are not events in the same way as virtual bodies are, since the virtual body eludes the dichotomy external-internal *in a much clearer way* compared to the so called real bodies. Because of its discrete and interactive nature, the virtual entity coincides with its story; it is a process. But it is not the mere sum of numerically different phases, for the weft of the body depends on interactivity. It takes place as an action making sense for a subject and from such interaction it acquires identity. This also means that it is a relative identity, dependent, and consequently fluctuant.

Certainly any body, if perceived by my body, is in a situation of interaction; nevertheless, it seems that being an object, and being an external object, carries the distinctive characteristic of not being able to be amended. This means that I cannot, *through a simple act of volition*, make an object not be so and so, make it not to be what it is: the external world would then be the world of the unamendable, to which all perceivable objects can belong—those in which we are now interested—but also the non-perceivable ones. From a theoretical point of view, the situation is different for a virtual body-image: even supposing that it is possible to separate a *simple volition* from a movement or a perception, considering that it will be possible, through sophisticated prostheses, to connect virtual bodies directly to the seats of nerve impulses, nothing will forbid to amend a virtual body through a simple act of volition. Once this is said, we have to ask ourselves if such an act is possible in its specificity only within the non-infinite conditions foreseen by the matrix, or if it is possible to implement algorithms allowing a retroaction on the matrix, that is to say a type of very strong interactivity, and if it is so, in what sense: a program that learns, modifies and develops through its relationship with the user. Given the interactive nature of the virtual, I do not see the theoretical impossibility of its happening, and so I do not see the impossibility of producing a kind of intersubjective communication mediated by the computer memory which would become, starting from a programmed base, a memory of experiences. If we ignore this possible development of the problem, still we have to consider that if the non-amendability is a characteristic which is necessary to objects belonging to the external world, then the virtual body does not belong to it. On the other hand, the virtual entity does not belong to the internal world: the object-event which it is, is not something I have imagined or dreamed of, but an environment which is surfable, by me or by other users, produced by a technology, and of such an environment I preserve the awareness of the difference from what I usually call "reality"—of which, as we have seen, a perfect reproduction is not possible. Finally, I would say that the virtual is neither internal nor external, since it is, if you wish, *external-internal*, considering that synthesis is not a sum, but something else, a proof of the ontological novelty of the virtual entity.

The virtual body-image is its story, the history of its phenomenalising in a series of relations which have been able to build a virtual environment including a human body, outfitted with certain prostheses. This leads to consider it as an object-event, which can be variedly interpreted, ontologically and with the related consequences: for instance, as a strange, relatively monotonous event allowing other bodies in its space-time; or as an object-event protracting itself in time according to a four-dimensional conception; or, in partially Spinozian terms, time being an entity of reason and the relationships among objects-events—not between objects and events—being one form of immanent causality, a succession of entities-instants. This last position is interesting, for in accordance with it, the permanence of the object in dynamism is a cognitive illusion. This leads to suppose that virtual bodies can be conceived as discrete not only in space, but also in time, as temporal segments which are numerically different, and that their diachronic identity is potentially discontinuous. As a result, the conceptualisation of a changing embodiment of the self is triggered, in order to grasp a performative form of self that can hardly be devised as a given set of faculties, and which is sensitive to the evolution of technology and programming languages. This entails the reassessment of the notion of self in terms of movement marks, based on its residual integrity as medium of transformations, with reference to its possible limits in the transitions between the actions defining the overall virtual environment.

15.5 Interactivity

Interactivity is an evident characteristic of the virtual (Kwastek 2013), which defines it among the other types of digital images: the virtual is an entity that phenomenalises itself in the interaction. In some ways, interactivity is something the virtual body-image has in common with any other body, but in some other ways it is a peculiar condition (Diodato 2012, pp. 53–67). To understand such peculiarity, and hereby to approach an ontology of the virtual, we must think about the concept of virtual and the difference between virtual and possible. As a matter of fact, in a general sense virtual is one of the states of reality, the revelatory of the hidden power of reality and, in even more general terms, the set "virtual" can no doubt be considered part of the set "real". The expression "virtual reality" seems indeed to pose no problem. Nevertheless, the concept of virtual can be better defined starting from the concept of "possible": while what is possible is conceivable as a constituted entity waiting to be carried into effect, what is virtual configures itself as a problematic complex, a junction of tendencies imposing an actualisation process. From this point of view, obviously the virtual-actual process does not correspond to the realisation process of what is possible, if this is conceived as a simple giving matter to a pre-existing form, and, in this sense, as the constitution of a substance, even though a dynamic one. Since the virtual environment develops itself in the interaction with the user, virtual means a dynamic configuration of forces with an intrinsic tendency to actualise themselves in forms which are not totally preconstituted (Heim 1993; Lévy 1995).

The virtual environment we are speaking of, together with all its perceivable qualities (the set of colours, sounds, tactile densities, etc.), that is to say, the environment in which I have the physical sensation of being immersed, is nothing but the actualising of the content of a digital memory, the staging of a binary-system algorithm. This leads to consider the relationship between *aesthesis* and *noesis*. We are actually facing the possibility of a reduction of *aesthesis* (as sensory perception) in computational terms, though this does not imply a reduction of the secondary qualities to the primary ones, and not even a possible reduction of the world to a number. We are speaking of an original and reversible solidarity between *aesthesis* and *noesis*, which expresses itself in an operative space of this sort: at one end there is a digital description in a computer memory, and at the other end a body outfitted with technological prostheses, with non-organic extensions of the senses. The body of the user of a virtual environment is a complex structure, a subject-object coming out of a technological project; it is a nearly-cyborg body, similar to the ones that were pondered and experimented by several artists, a body rendering itself in a phantomatic and eminently active entity. There is a lively discussion concerning these themes among the theoreticians of virtual reality (Hansen 2004, 2006; Heim 2014; Massumi 2014). In a certain sense, you can get to know a virtual environment only sensorially by an eminently bodily gaze, but at the same time a virtual environment is, as we said, a mathematisation of space, and its images are actualisations of algorithms. It is a paradoxical situation: the identity, the "self" of the user is at the same time de-corporised and iper-sensibilised: to meet a "thin" body, you have to be outfitted with a "heavy" body; the abilities of the organic body must then be technologically increased.

As it partakes in reality through the digitalisation process, which is circularly made possible by specific techniques, the ontological nature of the virtual should be studied even more in detail. Virtual entities, being constituted by the circular participation of the living body with the world, must not be understood as representations of reality, but as realities which are built in a way which is essentially different from the way others are. Thanks to the perception-vision, the world goes through the body and becomes an act, a movement of the body, which might be mediated by instruments of analogical reproduction, thus producing an image. Virtual bodies are instead "artificial windows giving access to an intermediary world" (Quéau 1989, p. 18). Now, in what sense is a "body" a window, meaning by this an internal-external passageway? The metaphor of the window is useful, provided one is to understand it in a non-trivial way: it is not about passing through Alberti's window, for the virtual environment is not—not only, and not essentially—a simulated reproduction of reality. The virtual entity is rather an environment-window, a *sui generis* place where the internal-external relationship changes according to the different standards, acquiring a relevant power. I will come back to this point from an ontological point of view. For the moment, let us refer to a famous quotation by Kandinskij: "Each phenomenon can be lived in two different ways. We cannot choose these ways, for they depend on the phenomena—they are derived from the nature of phenomena, from two of their properties: Internal-External" (Kandinskij 1978, p. 17). This is true not only of our own body, but also of everything that appears to us about the way our body shows itself: a phenomenon can be lived, somehow, from a certain distance; it can be perceived as other. It can be the world; however, the same phenomenon can also be a part of our life; it can engrave itself in it, accomplish itself as its *pathos* and show in this way, through visibility, its invisibility. All this corresponds to common experiences, which are just as selective as ordinary: something we perceive might engrave itself in memory and affectivity; it might become part of that primary and indemonstrable inside and might go back to the light of the common world through different means. Something else, the mass of the perceived phenomena, at least consciously, might not. But Kandinskij did not only claim that the phenomenon could be lived in two different ways, internal and external, but that this can happen because external or internal are properties *of* the phenomenon, of the same *phenomenon*. Since being at the same time internal and external belongs to its nature, the phenomenon can be lived as world or as *pathos*. Now, I do not know if this position can be held with what we generally call "reality", but it works well for virtual bodies: in a virtual body-environment, in which space itself is the result of an interaction, the world does not happen as if it was standing back or taking some distance. It happens in the way of a sense-feeling of immersion, and the body, since it is perceived as other, realises the sense of its reality, of its effectuality as a pathic and imaginary incision, as a production of emotion and desire, to the point where the sensation of reality transmitted by the virtual environment depends mainly on how efficiently the emotions are provoked in the user. From this point of view, precisely as reality, as something different from the user, as an environment in which it can interact, as bodies it can manipulate, virtual reality can provide its own, self-authenticating experience. So the virtual body-environment acts as an intermediary not only as mediacy between the computer model and the sensory image, but primarily between internal and external, the strange place where the border becomes territory.

15.6 Object/Event

As is well known, one of the most discussed questions in contemporary ontology is the distinction between object and event, and relatively the distinction between concrete and abstract. Now, in a virtual environment, what the user perceives as a thing is actually an event, the temporary actualisation of something virtual existing only, in its actuality, as an interactive relation function. This leads to the need for an articulated understanding of the concept of relation, as well as of the notions of thing and event as relational bonds, without this implying a drift, because the virtual has an actuality of its own, even beyond interaction—it is real exactly because it is virtual. What we have to do is to articulate, at least briefly, the question of the object-event relationship, in order to point out a typical ontological mark of the virtual body as it is here defined. Such a relation has been mainly conceived—if we decide, in order to ease our work, to overlook dialectic or neo-idealistic positions—as a form of relationship with two values: the event is an object (or more than one object) that changes. An ontology that admits events often conceives them as changes of an object, thing or substance provided with some permanence; so it conceptualises the event as relative to such becoming, even when the becoming object is not clearly identifiable. All in all, it is a conveyed of Aristotle's ontology, which conceives substance, with its intrinsic dynamism, as a main category. If we start from this assumption, we will have to face the question of the symmetry between event and object, and of the possible conceptual dependence of the category of event on that of the object, even if we conclude that the categories are not conceivable separately. Now, the virtual body-image, even though not amenable to a representation, does not exist as a body if not through interactivity. It *is* an interaction, an object-event, an action (or relation of interactivity) which is a body (i.e. virtual body), because it owns the characteristics we usually ascribe to bodies. As time goes by, the virtual body, if certain conditions concerning its interactive nature are given, persists to mutations of position, size, shape, colour, but also to certain conditions that concern its interactive nature. Virtual bodies are then—like, perhaps, bodies *simpliciter*—(relatively) monotonous events, but only if certain conditions are given. Pondering such conditions in the field we are interested in leads to transform questions like "do things like changes exist?" into questions such as "what are the conditions of possibility of the changes which are things?", and so we are brought to consider the typicality of the virtual body, leaving out for the moment the question about the ontological difference between the virtual bodies and the so called real bodies. In the case of a virtual entity, the event is something unrepeatable, a *concrete but thin individual* (meaning by this an integrated system), constituted by the interaction of a human body (i.e. a body-mind complex) outfitted with technological prostheses, with an electronic elaborator, implemented by an algorithm—in turn rendered by a programming language. Does this concrete but thin individual occupy exclusively one place? If so, what place? Several parts of my technological prostheses, several sensitive surfaces of my body, a certain part of my brain, a computer memory? In any case, it is a body which allows other bodies in its place, it allows, for instance, my body to pass through it, and if a virtual environment is a virtual body which is qualifiable as a structured surfable set of virtual bodies, then a virtual body can contain virtual bodies in its own body, like shadows, rays, angels, phantoms, and so on. Now, a virtual entity occupies a certain portion of the space-time, but not in an exclusive way, because the virtual body happens in the space-time of a non-virtual body; moreover, it multiplies its temporal form: what is its time? Of course it happens at the moment of the interaction, but among its conditions of possibility, or better, in its being a real body, there is the fact of being previously written or engraved on a material base, in a memory.

15.7 Ontological Hybrid

The virtual entity is, therefore, intermediary not only as it mediates between a digital model and a sensible image, but primarily as it is intermediary between interior and exterior as two sides of the same phenomenon. In this strange place the border turns into a territory. It can be concluded that virtual bodies-environments are neither simple images nor simple bodies, but bodies-images eluding the ontological distinction between "objects" and "events", since, like the "objects", they have a relative stability and remain the same through time, and, like the "events", they exist in the happening of interactivity. The ensuing individual is indeed solid, as it is the subject-object of actions and perceptions, but it is also "distinctively flimsy", precisely because of its being interactive. As a hybrid, its ontological status is uncertain; one might even call it a flimsy body in a non-continuous world, made of points-data coming across in terms of fluidity and density to saturate perception. Digitalisation makes this body light and interactivity makes the condition of its appearing. Such a body finally belongs to the natural and artificial worlds, perfectly in line with both theoretical possibilities famously outlined by Aristotle's very powerful and paradigmatic definition of existing things in the *Physics*: "Among beings, some are formed by nature, some by other causes. Among those formed by nature, we may name animals and their parts, plants, and the simple bodies (earth, fire, air, and water); all of these, together with beings like them, we call 'formed by nature'. Observation disclosed how they differ from things not constituted by nature: each of them has within itself a beginning of movement and rest, whether the 'movement' is a local motion, growth or decline, or a qualitative change. Such is not the case with things like beds and clothes: that is to say, to the extent that these come within the classification of ‘products of art’, they do not have implanted within themselves any tendency to change; nevertheless, in so far as they happen to consist of stone or earth or a composite material, they do have such a beginning of movement and rest, but only in this respect. But even this circumstance gives evidence that the nature of a thing is in some sense the factor which initiates movement and rest within that thing in which it is itself immediately, not incidentally, present” (Aristotle 1961, p. 23; *Physics*, II, 1, 192 b). Now, virtual entities, as digital appearances existing only in interactivity relation, are clearly artificial, a product of art, however they also have an “innate” impulse to change, which does not depend on their “natural” components but rather on their own “nature” of event. Hybrid then means artificial-natural, in some respects almost a “living” system, or, in the current jargon of physics, a dissipative system. Not only an *object-event*, then, but also a *subject-object*, virtual bodies only exist—as previously remarked—as the encounter between digital writing and bodies that are sensitive to it, hence as constitutive interactivity. This allows to conceive of the relation (i.e. the encounter) as in itself capable of constituting entities—independently of specific relational properties—hence to develop an ontology of relations—until now mostly uncharted—, as a grateful addition to the world’s furnishings.

The ontological event I have tried to track down here, a crossroads of multiple relations in the making—bodies are set of events and not material beings which occupy a given volume of their own that is the same for all observers—is neither physical nor psychical, but rather simultaneously virtual and real, the plexus of the perceiving, the perceived, and their relation, a gesture that is a thought. This is the aesthetic *logos*, the chiasm or embodiment of *logos*. Events in turn produce complex intersections intertwining with other events, and integrating events as parts. They can then play the role of subjects and reversibly of objects, whenever “seized” as a part of other events. Subject, object, active, passive are then reversible processes functions. In other words, the subject is no longer a *subjectum* or *substratum*, but a provisional point-place of some space-time phases in the process. Starting from such an ontological situation, we can reassess some entities which make display of an essentially relational status. The virtual break the loop connecting *physis* and *techne*, confronting us with a form of *poiesis* which requires a new thinking approach. Who or what does produce the “things” we see and perceive as if they were the product of a creating will? What power does create here the dimension of aesthetic experience, clearly the most complex experience for human beings? Which relationship between freedom and necessity is here staged? The actual convergence of freedom and necessity, that is to say, the power of the imagination and of the technological apparatus, inasmuch as it is an incompatible double production, does not comply to rational accounts, but it is rather simply given, at times also as an artistic event. What is at stake is the being, the things, the world, and this is what the human and sometimes super-human action of art displays in an exemplary way. In the virtual body-images this convergence or productive power is consciously sought in the genesis of the image. The digital process allows the encounter between the extreme contingency of the personal gesture and the anonymous and petrifying gaze of the technological process, in other words the transformation of reality into number, the breaking down of complexity in a generally computable complication.

I think that the ontology of the virtual entity reveals how fundamental is the category of relation—not the relationships between already constituted polarities, but rather the relation as something that is constitutive of polarities.

References

Aristotle. (1961). *Physics* (R. Hope, Trans.). Lincoln: University of Nebraska Press.

Deutsch, D. (1997). *The fabric of reality: The science of parallel universes and its implications*. New York: Penguin.

Diodato, R. (2010). Il corpo virtuale come esempio di sistema. In L. Urbani Ulivi (a cura di), *Strutture di mondo. Il pensiero sistemico come specchio di una realtà complessa* (pp. 249–269). Bologna: Il Mulino.

Diodato, R. (2012). In S. Benso (Ed.), *Aesthetic of the virtual*. Albany: Suny Press.

Diodato, R. (2013a) *Relazione e virtualità. Un esercizio del pensiero estetico*. Bologna: Dehoniane.

Diodato, R. (2013b). The virtual body. In B. Antomarini & A. Berg (Eds.), *Aesthetics in present future. The arts and the technological horizon* (pp. 9–23). New York: Lexington.

Diodato, R. (2014). About virtual experience. *Metodo. International Studies in Phenomenology and Philosophy, 2*, 47–68.

Diodato, R. (2015). *The sensible invisible. Itineraries in aesthetic ontology*. Szczecin: Mimesis International.

Diodato, R. (2018). Phenomenology of the virtual body: An introduction. *Analecta Husserliana. The Yearbook of Phenomenological Research, CXXI*, 568–579.

Grau, O. (2003). *Virtual art. From illusion to immersion*. Cambridge: MIT Press.

Grimshaw, M. (2014). In M. Grimshaw (Ed.), *The Oxford handbook of virtuality*. Oxford-New York: Oxford University Press.

Heim, M. (1993). *The metaphysics of virtual reality*. Oxford: Oxford University Press.

Hansen, M. (2004). *New philosophy for new media*. Cambridge: MIT Press.

Hansen, M. (2006). *Bodies in code: Interfaces with digital media*. New York: Routledge.

Heim, M. (2014). The paradox of virtuality. In M. Grimshaw (Ed.), *The Oxford handbook of virtuality* (pp. 111–125). Oxford: Oxford University Press.

Kandinskij, W. (1978). *Point and line to plane*. New York: Dover.

Lévy, P. (1995). *Qu'est-ce que le virtuel?* Paris: Éditions La Découverte.

Kwastek, K. (2013). *Aesthetic of interaction in digital art*. Cambridge: MIT Press.

Lopez, D. M. (2010). *The philosophy of computer art*. London: Routledge.

Massumi, B. (2014). Envisioning the virtual. In M. Grimshaw (Ed.), *The Oxford handbook of virtuality* (pp. 55–69). Oxford: Oxford University Press.

Quéau, P. (1989). *Metaxu. Théorie de l'art intermédiaire*. Seyssel: Champ Vallon/INA.

Sartre, J.-P. (1986). *L'imaginaire. Psycologie phénoménologique de l'imagination*. Paris: Gallimard.

Urbani Ulivi, L. (2018). Mind and body. Whose? Philosophy of mind and the systemic approach. In L. U. Ulivi (Ed.), *The systemic turn in human and natural sciences. A rock in the pond*. Cham: Springer.

Chapter 16
Techno-Aesthetics and Forms of the Imagination

Pietro Montani

Abstract The concept of "techno-aesthetics" refers to the fact that human sensitivity (*aisthesis*) is commonly exercised via technical extensions of the body. I interpret this phenomenon here against the backdrop of a philosophical paradigm presented by Immanuel Kant in his *Critique of the Power of Judgment* (1.1.). As well as an absolutely central role played by the quality and performances of sensitivity, this paradigm includes a "realistic" and embodied conception of the imagination (1.2.) and a techno-aesthetic approach to language and linguistic creativity (1.3.). In both cases, the emphasis will be placed on the convergence and differences between the Kantian approach and the phenomenological orientation linked to influential theoretical models such as Embodied Cognition and the Material Engagement Theory. The second part addresses the specific issue of art. More explicitly, I shall ask in what way art can constitute an epistemic object for a techno-aesthetic (2.1.) and whether such a theoretical classification of art does not result in the assumption of an element of historical variability (2.2.) which prompts us to supplement Kant's formulation. Indeed, the latter could not respond satisfactorily to the, sometimes radical, cultural changes attributable to technological innovation and often highlighted by the arts. The conclusions identify three orientations of that highlighted which are "experimental", "revelatory" and "deconstructive".

Keywords Technics · Sensitivity · Embodied Imagination · Arts

P. Montani (✉)
Sapienza Università di Roma, Rome, Italy
e-mail: pietro.montani@uniroma1.it

S. Chiodo, V. Schiaffonati (eds.), *Italian Philosophy of Technology*, Philosophy of Engineering and Technology 35,
https://doi.org/10.1007/978-3-030-54522-2_16

16.1 The Critical Meaning of Techno-aesthetics

16.1.1 "Technical Schematisation"

The expression "techno-aesthetics" appears in a short text by Gilbert Simondon (2014), a letter dated 1982 to Jacques Derrida but never sent. In it Simondon wished for the theoretical foundation of an aesthetic approach to technical practices and indicated some of their potential directions. Although the direction taken by this article is, in many ways, indebted to Simondon's philosophy of technology (Montani 2007, 2014), I shall here speak of "techno-aesthetics" in a very different sense. I intend to show that the link between the two words forming the expression can be explained most congruously within the context of Kant's critical philosophy (which has no significant correlations to Simondon's thought). More specifically, the fact that human sensitivity (*aisthesis*) is also exercised via technical extensions of the body—and as in a celebrated image is also active "at the tip of the blind man's stick" (Bateson 1973; Malafouris 2013)—will be interpreted here against the backdrop of a philosophical paradigm which, like that presented by Kant in his aesthetics (Kant 2000a), underlines the absolute importance of the quality and performances of sensitivity itself for the purpose of a proper critical understanding of the human experience and its distinguishing requisites (Garroni 1986, 1992; Nanay 2016). Allow me to add that my personal journey towards a philosophy of technology has taken shape in the course of habitual reading of Kant's texts, wherein the issue is decisively, albeit not explicitly, present. In the first part of this article, I shall discuss the salient theoretical points; in the second, I shall propose some of the developments linked to the arts.

How does the sensitivity (*Sinnlichkeit*, *Gefühl*) explored by Kant (2000a) meet the needs of a philosophy of technology? The answer is very simple. Distancing itself massively from the Cartesian, representative and self-centred subject, the Kantian aesthetics refers to a subject that we could describe as ek-static. A subject that "feels" the world-environment and "feels himself/herself" in the world-environment in the terms of an essential co-belonging (material and to a community). The performance which enables this subject to process his/her ek-static condition in an experience proper (e.g. in a shared cognitive experience) forms part of the concept of "schematism", used by Kant (1998, 2000a) to define the construction of the mediations required to guarantee the human experience has a constant and interactive relationship with the world-environment. The key point here is that, when describing these mediations, Kant shows that he also, and perhaps above all, sees them as embodied technical mediations. That is, true sensorimotor schemas engaged in a technique or willing to extend themselves in technical artefacts, including the particularly powerful technical artefact that is language (see Sect. 16.1.3). Already in the first *Critique*, the "schematism" had been explicitly presented as the product of a *Kunst*, i.e. of a technique (Kant 1998, p. 273). I would now like to better justify this theory, which highlights an unusual profile of Kant's philosophy.

The specific relevance of the technique appears very prominently in the first (unpublished) introduction to the *Critique of the Power of Judgment* (Kant 2000b). Kant is searching for the principle which the activity of judging in its most original form obeys; he defines this activity as the reflecting power of judgment, when, i.e., there is a single case and the power of judgment must find, or construct, the rule under which it is to be subsumed. It is a classical example of the hypothetical inference of an empirical rule based on a set of data and observations. It is also, *and more originally*, a classical example of the production of an artefact, e.g. a chipped flint. How does the schematism of the power of judgment behave in such circumstances? Here is Kant's remarkable reply:

The reflecting power of judgment thus proceeds with given appearances, in order to bring them under empirical concepts of determinate natural things, not schematically, but *technically*, not as if it were merely mechanically, like an instrument, but *artistically*, in accordance with the general but at the same time indeterminate principle of a purposive arrangement of nature in a system, as it were for the benefit of our power of judgment… Thus the [reflecting] power of judgment itself makes the *technique of nature* into the principle of its reflection *a priori*. (Kant 2000b, p. 17)

If the idea of a "technique of nature" must be considered a "principle *a priori*" of the reflecting power of judgment, this means that the conditions of the human experience include *first and foremost* a widespread perception of a world-environment that shows an indeterminate (but determinable) readiness to respond to technical inspections. However, it must be clear that this is not a logical principle but an aesthetic principle (a feeling and a feeling oneself). Moreover, it is not an objective principle (a conceptual definition of "nature") but simply an indeterminate tendency to perceive the world-environment and interact with it in terms of technical virtuality. "As if", says Kant, nature were an artefact constructed on the basis of rules favouring our ability to judge. Finally, this principle is not limited to the production of conceptual schemas because its most original nature is inherent (as we shall see more clearly below) to a sensorimotor context. In other words, the expression "*a priori*" must not suggest to us some "innate" device that guarantees the systematic nature and consistency of our (cognitive and practical-operational) experience of the world-environment. Using different language from Kant's but remaining true to the paradigm of critical philosophy, we might say that we human beings feel the world-environment as experiential terrain tuned *first of all* into a technical register. We feel it as a world that comes towards us, tuned into our need to find an abode in it and encouraging us to seek its technical mode. This is how our sensitivity is able to extend itself into technical artefacts and incorporate technical artefacts, such as the stick mentioned at the beginning but also the entire, complex and non-, family of prostheses with which the agency of the body spontaneously identifies. Finally, for the same reason, our experience is not equipped with any automatic execution programme; indeed, it is strongly dependent on the manner in which the "technical harmony" I have just mentioned rearticulates and reorganises itself over time. I shall add that, as regards the specific importance to attribute to this time and transformative factor, a philosophy of technology must quite radically reorganise the Kantian premises I am discussing (see Sect. 16.2.2). In the final introduction to the third

Critique, Kant reformulated the principle of the "technique of nature", defining it more generically as a "purposiveness of nature", but this reformulation changes nothing in the essence, as I shall demonstrate below with regard to the specific case of language. Firstly, however, I would like to stress a point that has already emerged informally but warrants a more focused thematisation. I am referring to what I would like to call the "realism" of the Kantian imagination.

16.1.2 Realism of the Imagination

Although Kant hesitated at length to attribute the management of the schematism to the imagination (*Einbildungskraft*) rather than to the understanding (*Verstand*), in the third *Critique* he explains that the emergence of the empirical schemas occurs in the exercise of "free play" between the two faculties. A free play that the aesthetic judgment exhibits in an exemplary manner as an agreement that occurs spontaneously and effortlessly. It is the imagination, however, that opens and drives this game in which the intellect operates in the mode of a mere and indeterminate "lawfulness". In the free "schematism of the reflecting power of judgment", writes Kant (2000a, p. 167), the imagination "schematises without any concept", so that the judgment itself "must rest on a mere sensation of the reciprocally animating imagination in its freedom and the understanding with its lawfulness" (*ibid.*).

But how might we portray the specific performance of the imagination? It is certainly a synthetic operation, i.e. the addition of regularity to the mere "synoptic" collection of the empirical multiple. But to which regularities are we alluding if they are not conceptual regularities in the strict sense? The answer lies in the sphere of technical performances linked to the "free play" Kant speaks of. The resulting frame is that of an imagination constantly interacting with the "things" in the world-environment. The Kantian text asks us to see this interaction as a toing and froing between attentional moments focused on certain material affordances and moments of indeterminate and diffuse attention (Nanay 2016); between a listening focused on the main properties of the empirical fact and one centred on the virtual and, so to say, "supervenient" ones (Montani 2014). It is during this toing and froing that a determinate project appears, on which the intellect will eventually impose its power of unification. We must not, however, only think of a conceptual determination in the strict sense, such as the hypothetical formulation of an empirical law translatable into a linguistic enunciation. We must also, and indeed primarily, think of a determination that is manifested in the form of the overall agency of a body engaged in active production, starting from the hands which, in this frame, eminently represent the fundamental embodiment of a performance that I describe here in terms similar to those used in the "Material Engagement Theory" (Malafouris 2013). But with one difference, namely that the "material engagement" does not bother to distinguish, within its device, the presence of *two different instances* (key to Kant's critical philosophy) which are ascribed differentiated tasks (although this difference is sometimes simply recorded; see Malafouris 2013, pp. 99–106). In other words, following and clarifying the text of Kant's aesthetics permits us, in the free schematism of the power of judgment, to identify a synergy between the *freedom* of imagination and the *lawfulness* of the intellect, the first sphere of practice of which presents the characteristics of technicity or, if you prefer, technically invested knowledge.

I shall return to this shortly, but I wish here to stress one of its essential points. As well as its reproductive nature, i.e. the ability to conserve a trace of previous percepts and represent them in their absence, we must also include its essentially *interactive* nature among the principal requisites of the Kantian imagination; i.e. a specific sensitivity to the affordances highlighted by the world-environment and those of a technical nature in particular. Before being a form of *phantasia* (to use the Greek word that thematises the ability of the imagination to re-present something in its absence), Kant's *Einbildungskraft* is an imagining reality in the sense that it interacts with the world-environment and shapes it just as much as it is shaped by it. That the objective affordances of the world-environment help "shape" the imagination during a technical activity is certified by Kant in an extremely refined passage of the *Critique of the Power of Judgment* (Kant 2000a, § 21); it says that, in the process of empirical knowledge in the strict sense, the imagination has the task of defining the free play with the intellect based on a "proportion" linked to the purposiveness of the object we wish to know. It is the object of knowledge, therefore, that exercises feedback on the freedom of imagination, steering its interactivity towards one certain conformation rather than another. Here, Kant is speaking of conceptual knowledge, but it is quite clear that the interactive situation defined by the idea of a "proportioning" prompted by the affordances of a determinate object can be grasped in a more original manner in technical practice. To return to our stick, simply think of how many other uses it can be put to, depending on whether the focus is on certain of its qualities rather than others. For instance, on the flexibility required to design a bow for shooting arrows or, conversely, on the resistance required for a lever, etc.

We note, lastly, that the interactive and "realistic" performance I am describing explores a broad field of determinations, the nature of which can be linked to *semantics* rooted in the embodied and technical nature of the imagination (I shall return to this shortly). Moreover, the sensorimotor schemas that *emerge* during this interaction make themselves available to a broad and flexible use that Kant did not fail to discuss in one of the most important and innovative passages of his aesthetics.

Before addressing this, I must explain a certain spilling over of my argument onto the terrain of anthropology and cognitive archaeology (to which the theory of "material engagement" belongs), something that might be considered incompatible with Kant's paradigm. On the contrary (Montani 2017), I am convinced that I am not taking a liberty by linking some aspects of critical philosophy to the anthropological sphere because it helps us reconsider its methodological utility and relevance today, as we shall see now in light of a particularly fecund Kantian suggestion. By contrast, it is important to define clearly the point at which the argument must indeed distance itself from some distinctive traits of Kant's philosophy (but not from the manner of thinking that distinguishes a critical philosophy).

16.1.3 Imagination, Technics and Language

Kant is, therefore, very clear that the imagination does not submit only "blind syntheses" to the intellect (as stated in the first *Critique*) in "free play" but true hypothetical schemas that the intellect can then synthesise in its own way, i.e. with a conceptual rule. This can, of course, be a conceptual rule *linguistically explainable* in an empirical law true and proper (e.g. the law that defines the orbit of the planets based on the schema of the ellipse). But, as we have seen, it may also be a cognitive rule incorporated into an artefact, i.e. *a rule of art in the broad sense* (art as *techne*, e.g. a tool such as a bow for shooting arrows) and in the specific sense (so-called fine art or work of genius, see Sect. 16.2.2).

That these are rules, i.e. synthetic events (albeit not intellectually determined), is demonstrated firstly by the fact that an incorporated rule can be *exported* to other spheres, as a constructional principle for the creation of other artefacts. The synthetic schema of the bow, for instance, i.e. the *embodied* rule that enables me to construct bows of varying sizes with different flexible materials etc., can be transferred to the design of a bridge arch or the vaulted ceiling of a building. In this "exportation" we can fully grasp the play between what pertains to the sensorimotor component (I can teach someone to make a bow for arrows without making any use of language) and what pertains to the linguistic-intellectual component. By adopting the same rule to construct a totally different artefact—a bridge rather than a bow for shooting arrows—what the artefice conserves is only the bow's ability to charge itself with a force and not also that of converting this force into a ballistic performance. It is evident that, in this transfer of an *embodied* rule from one type of artefact to another, the mode of sensorimotor teaching reveals its limitations and we enter into the context of a decisive relationship with language. This corroborates Kant's methodological principle which consists in keeping the two components of "free play" spoken of above separate.

There is an important passage in the *Critique of the Power of Judgment* that explains and formalises this topic. I am referring to § 59, wherein we can say Kant conducts the whole work of rethinking technical schematism to reach one of his most important results: the idea of a particular—and specifically creative—form of schematism, which he describes as "symbolic exhibition".

In the paragraph in question, Kant identifies two manners of exhibition (or "hypotyposis") in the work of the imagination. The first is schematic in the objective sense of the word: the imagination provides the sensorimotor schema required to guarantee the empirical concepts have a stable and direct semantic reference (see Sect. 16.1.2). The second, by contrast, exploits the imagination's aptitude to work not only directly (objectively, referentially) but also indirectly and analogically, and is called "symbolic".[1] The exhibition is called symbolic when an "idea" (i.e. a concept

[1] This is a most unfortunate definition, given the great variety of meanings attributed to the word "symbol" in various theoretical and philosophical contexts. Kant would have done better to speak of "analogical exhibition".

of reason for which no direct sensitive intuition would be adequate) is provided with an exhibition to which regard the procedure of the power of judgment is "only analogous" to that which occurs in an objective schematism: "i.e.", Kant writes, "it is merely the *rule of its procedure*, not of the intuition itself, and thus merely the form of the reflection, not the content" (Kant 2000a, p. 226, my italics). The definition, expressed in a slightly contorted manner, is clarified shortly afterwards by a particularly good example:

> Thus a monarchical state is represented by a body with a soul if it is ruled in accordance with law internal to the people, but by a mere machine (like a handmill) if it is ruled by a single absolute will, but in both cases it is represented only *symbolically*. For between a despotic state and a handmill there is, of course, no similarity, but there is one between the *rule for reflecting on both and* their causality. (Kant 2000a, p. 226, my italics)

Kant adds that our *language* is full of these indirect exhibitions in which "the expression does not contain the actual schema for the concept but only a symbol for reflection": "Examples are the words *ground* (support, basis), *depend* (be held from above), from which *flow* (instead of follow), *substance* (as Locke expresses it, the bearer of accidents), and innumerable other non-schematic but symbolic hypotyposes". (Kant 2000a, p. 226)

Kant's almost random examples are highly significant for the theory I am developing because, to varying degrees, they show the *somatic rooting* of this indirect and analogical mode of schematism, as if the imagination had to attach itself firstly to a typical corporal and sensorimotor experience. Ground (*Grund*) is what connects to the act of gaining a firm base for our feet (and for an erect stance); depend (*abhängen*) is being and feeling yourself hung from something that might give way (so almost the opposite of *Grund*); flow (*Fliessen*) is the act of following up but also the feeling of drifting, a body flowing in the current; substance (*Substanz*) is the substrate of accidents but also and primarily the physical act of sustaining something (we remain in the constellation of the grounding and depending).[2]

In this way, we realise that the example of the monarchical state also responded to the same requisite since the "form of the reflection" here was provided, yet again, by a *body* (and, to use a famous Kantian image, we could say that each of its organs is both means and end) or by a *bodily action* (turning a millstone with the arm, or hand [*Handmühle*]).

The nature of the intuitive reference is therefore—and systematically in the examples made by Kant—a bodily experience and *a somatic aptitude to schematisation*. But, if this is so, what "rule" ought to allow the analogical exportation that Kant seeks to illuminate? Here, it really seems that we must look to the *technical schematism* mentioned in the first introduction. In other words, an understanding of the semantic performances executed by language by virtue of an analogical procedure seems spontaneously oriented towards *the relationship between the human*

[2] To fully appreciate the huge epistemological importance of this point we must stress the fact that "substance" is indeed a pure concept of the understanding (i.e. a category), which Kant here connects not only to the common usage of language but also to the embodied reference that lends meaning to language.

body and the production of an artefact, or an artefact into which the human body extends itself and its *aisthesis*. That is to say, if language can benefit from a "rule for reflecting", it means that this rule appears firstly as a synthetic result of a technical schematism (or "material engagement", if you prefer). In turn, and as I have already mentioned above, the extension of the rule to other artefacts rests on the guarantee, *which only language can offer* (or that language seems designed to offer), of being able to dissociate from a direct intuitive exhibition and shift analogically to another intuition. More precisely, it is the *embodied* and externalised schematism of the imagination that produces these shifts, but it is language that, in the ultimate analysis, guarantees its formalisation proper. We are therefore dealing with a cognitive *blending* (Fauconnier 1997; Fauconnier and Turner 2002), but Kant believes the two "mental spaces" coordinated by the analogical exhibition belong to different fields: one has to do with the freedom of imagination and the other with compliance with laws of the intellect (gathered and formalised by language).

In all these cases, therefore, Kant sanctions us to imagine a *technical schematism* into the performance of which the contribution of an artefact, seen as an extension or part of the body, can easily be incorporated. Let us imagine, for *Grund*, a golfer "taking his/her stance" with both feet to inject the right force and precision into a stroke hit with the club; a climber tackling a risky passage who becomes one with the harness from which he/she, literally, "depends" (*abhängen*); a little girl helter-skeltering down a waterslide on an inflatable cushion (*fliessen*); or a labourer engaging his/her whole body to support a weight, e.g. two brick-loads hanging from a plank which calls for constant rebalancing (*Substanz*).

What is at play is *a specifically techno-aesthetic mode of schematism*, which motivates the semiotic device underpinning the analogical exhibition. This activity also forges a particularly productive relationship with language as it allows the linguistic experience to extend to the point that the ideas of the reason become sensitive—indirectly but still linked to a sensorimotor context.

In other words, the "rule for reflecting" Kant speaks of is one that *emerges* during the schematic and synthetic operation that allows a technical tool (e.g. something like a hand-mill or the entire category of artefacts founded on the same sensorimotor schema) to be imagined, produced and utilised. It is therefore one of the *embodied* rules I spoke of earlier and we can now better see its link with an action of "remodelling" the imagination. *A rule of technical operation and not a conceptualising one*. A rule for which reference to an "aesthetic" in the sense in the first *Critique* is not sufficient; we must adopt a techno-aesthetic—an aptitude to feel our "own body" extended in a prosthesis. That is to say a fundamental syntony of the imagination with the identification of technical affordances in the world-environment.

To summarize, it is thanks to an original techno-aesthetic that the technical schematism managed directly or analogically by the imagination can perform a semantic function of primary importance, arriving to cover (as Kant suggests) a very broad sphere of the linguistic significations and syntactic articulations most closely bound to the agency of a body. From the specific perspective of a techno-aesthetic, what I have just argued is strongly validated by a series of theoretical approaches (only partially unifiable) linked to the crossover area that for about the last 30 years has formed among the philosophical tradition (particularly that of phenomenological origin), the neurosciences, palaeoanthropology and cognitive archaeology. I am thinking, in particular, of works such as Turner (1996); Fauconnier (1997); Lakoff and Johnson (1980); Tomasello (1999); Gallese and Lakoff (2005); Gallagher (2005, 2017); Malafouris (2013); Ihde and Malafouris (2018), but also, in terms of rethinking a critical philosophy, Ricoeur (1975); Garroni (2005); Montani (2014, 2017); Malabou (2016). The advantage of the Kantian perspective is that it paves a very promising way to addressing the extremely controversial issue of the relationship between embodied cognition and language (Montani 2018).

16.2 Techno-Aesthetics and Art

16.2.1 Art in General, Fine Art and Innovative Art

In the passages Kant (2000a, Paragraphs 43–54) devotes to art (generically or with regard to the products of so-called fine art), the issue of *technical schematisation*—i.e. producing rules *incorporated* into an artefact—is not only well established but also specifically developed. I would like to address this point starting from an aspect I have thus far intentionally kept in the background, simply touching upon it rhapsodically in some passages wherein I have referred to the "quality and performances" of human sensitivity (*Sinnlichkeit*) without differentiating between its precise form (*Empfindung*, a defined sensation) and its broad form (*Gefühl*, an indeterminate sentiment or sense, *Sinn*) as, conversely, Kant does repeatedly. Both aspects appear in the critical discussion of sensitivity, which is at the core of the third *Critique*, and both are correlated to the human body and its specific ek-static agency. It is, however, in the sphere of the *Gefühl-Sinn* that the *aesthetic principle* of the power of judgment is originally formed. That is to say, the indeterminate sentiment of a "technique of nature" (first introduction) or of a "purposiveness of nature" (final text) as it is a basic mood (*Grundstimmung*) of our being in the world, as we might say in Heideggerian terminology (Heidegger 2010), perfectly befitting the situation. This is an original "feeling of pleasure *and* displeasure" (one *and* the other, as we shall see now), argues Kant, that "synchs" the way we feel ourselves in a meaningful relationship with the world-environment.

Kant explains very clearly that this sentiment is none other than the effect of the "free play" of imagination and intellect. That is to say, and as we have already said without pursuing the point, it is the primordial form of schematism (and its intimate technical values), a form that is exemplarily manifested in the aesthetic judgment: "Since the freedom of imagination consists precisely in the fact that it schematizes without a concept, the judgment of taste must rest on a mere sensation of the reciprocally animating imagination in its freedom and the understanding with its lawfulness" (Kant 2000a, p. 167).

To free the field of the suspicion that the topic just touched on is an abstract and speculative one, let me simply indicate its congruency with the latest neuroscientific and psychobiological studies in which the pertinence of the emotions in determining the conducts, adaptative ones included, of the living being has been extensively ascertained and experimentally corroborated. In particular, Panksepp and Biven (2012) showed that, among the *basic affections* which the subcortical region of the brain seems to possess, "seeking" figures as an *affection* that is to some degree *superordinate* to the others (fear, rage, lust, care, panic and play) in the emergence of the ancestral forms of the "mind". The reason is clear: it is a widespread sensation of being able to find an order which helps the biological resources of the living being that gives the latter the drive required for life and its unpredictable virtualities. After all, the life instincts that Freud (1990) subsumed in the concept of "Eros" say the same thing: that life knows how to forge a path even in the midst of constraints and seeming disorder (*Eros* was the son of *Poros*, expediency, and *Penia*, poverty). Now, Kant's reasoning is perfectly compatible with this way of posing the problem: the "sentiment of pleasure" that synchs the ek-static condition of the human being is nought but the most original emotion linked to its engagement in the forms of crucial interactivity with the world-environment that I linked to technical schematism and the realism of the imagination in the first part of this article. The fact that the human being has converted this favourable feeling into its distinctive *technical creativity* (Montani 2017) reaffirms the fact that the Kantian concept of a "technique of nature" is equipped with an empirical adequacy superior to that of the vaguer "purposiveness of nature".

Two things remain to be explained: what is the role played in this framework by the "pleasure-displeasure" relationship evoked above and how might art help clarify this point?

In reality, the issue of art "in general"—i.e. that of *Kunst* as a *techne*—is already analytically contained in my discussion in the first paragraph. I have constantly linked "technical schematism" to an embodied practice, not only in the sense that it is executed physically (it is a "work of the hand", to use another locution dear to Heidegger) but also in the sense that, in its constitution, it brings in the material artefacts that emerge from the interactive context in which it takes place (Malafouris 2013; Parisi 2019). To this end, Kant (2000a, Paragraph 43) differentiates between the automatic execution of a programme, typical of artefacts reliant on instinctual protocols (e.g. the hive) and the production via "rational deliberation" typical of human technics. His incidental, but significant, example is the manufacture of a pair of shoes. So, how can we differentiate a pair of shoes from a product of "fine art"? Kant's answer is remarkable: in fine art, he observes, there is a specific "animation" (*Begeisterung*) which assimilates it with "nature", seen here as productive and autopoietic spontaneity. Kant (2000a, Paragraph 45–50) gives the name "genius" to the specific "proportion" of power that presides over the embodiment of this type of *originally* productive technical schematism. And it is, of course, a proportion in which the imagination (and, again, we must keep a focus on its realism) earns absolute pre-eminence. This, therefore, is how the work of genius in art becomes significant for a critical philosophy: it places the productive imagination's aptitude to interact with the world-environment in such a way as to generate *new artefacts* before our eyes, without ever settling onto a permanently stable plane. In other words, the work of genius in art gives us an experience of the creativity via which the human being's material engagement legitimises itself, constantly modifying previous balances and introducing *new rules* into them. Here we reach the cut-off point of Kant's reflection. A point that contains an opening towards the historical-temporal issue evoked above (see Sect. 16.1.1), but only thematises it in a tangential and reticent manner. It is the same point at which we see the emergence of a fitting critical understanding of the issue of "displeasure", or of the negative and discordant element that arises in the aesthetic principle of the power of judgment. I shall now try to explain these points better.

16.2.2 The Exemplarity of the Work of Art. Technical Innovation and the Institution of Historical Communities

In speaking as I just have of a "material engagement" of the human being who legitimises itself by *modifying* the previous balances and introducing new rules into them, I have simply reformulated with different vocabulary what Kant says very clearly regarding the "properties" of genius. I shall quote and comment on the first two:

(1) Genius is a *talent* for producing that for which no definite rule can be given; it is not a mere aptitude for what can be learnt by a rule. Hence *originality* must be its first property. (2) But since it also can produce original nonsense, its products must be models, i.e. *exemplary*; and they consequently ought not to spring from imitation, but must serve as a standard or rule of judgment for others. (Kant 2000a, p. 188).

The first property of genius, as we see, highlights the *radically original* aspect of its production. It is a production that cannot be learnt from a rule for the precise fact that it consists primarily in *bringing out* a new rule. But, because a new rule would not make sense (it would be a *Unsinn*, Kant writes) unless it were also adopted by a community as a *model*, the second property of the work of genius—its "exemplarity"—implements an overall reorganisation of the forms of community life that are constituted following its appearance.

In the (extreme) formulation I have just proposed, I have deliberately forced Kant's text, drawing it out of its reticence. It is obvious, nonetheless, that the "others" alluded to in the text are the members of a community of which we shall be able to speak *only in the future perfect tense*, i.e. the community which "will have formed" in the space opened up by the innovation "implemented" by the productive imagination (and here we have another fully warranted Heideggerian assonance).

Reshaping the existing forms of life, reforming or re-evoking the already given rules means that the work of genius introduces instability and uncertainty into the world-environment and this too forms part of its exemplarity.

Not only is this confirmation of the "sentiment of life" (*Lebensgefühl*), it is also a form of *disorientation*: pleasure *and* displeasure. The *Grundstimmung* which tunes into the aesthetic condition of the human experience, therefore, is a feeling of being in profound and harmonious contact with the world-environment but also, *and at the same time*, a feeling of being uprooted and even extraneous to it. It is for this very reason that the technical schematism we have spoken of never settles itself into a permanent model and is distinguished by its infinite innovativeness.

As is therefore clear, in the parts of the text on the aptitude of the human experience to extend and reorganise itself, Kant reached the threshold at which the intimate *historicity* of the ek-static subject—of which the third *Critique* describes the original (techno)aesthetic foundation—starts to appear. It is a threshold that critical philosophy—when seen predominantly as transcendental philosophy—is not required to cross. Others after Kant have crossed it, remaining, some more and some less, true to his line: Schiller, Nietzsche and Heidegger (Montani 2007). Nonetheless, on this same threshold, techno-aesthetics still has something to say regarding the specific issue of art. I shall end by indicating three ways of articulating the issue. One "experimental", one "revelatory" and one "deconstructive".

In what way, we must ask ourselves, would technical innovation introduce "changes" into existing life forms? Clearly, the first level of pertinence of this question is linked to the changes that a reorganisation of the technical balances might bring to the sphere of human sensitivity. Has the *aisthesis* of human beings changed throughout history in relation to technical reorganisations of the world-environment they have lived in? We must give this question a cautiously affirmative answer while taking care, nonetheless, to add that it is not immediately apparent on which level a suitable theoretical treatment should be placed. Some disciplines I have already referred to, such as cognitive archaeology and palaeoanthropology, the neurosciences and psychobiology, are concerned with the issue, but it would be rash to conclude that their respective results might be unified in a homogenous interdisciplinary model within the scope of which one could give a reliable reply to the specific question on changes to sensitivity. For example, the fact—now extensively certified at experimental levels—that the acquisition of new skills (e.g. think of someone learning to play a musical instrument) implicates significant neuroplastic changes could certainly not be extended to the "aesthetic" plane as has been discussed here with regard to the Kantian approach. Moreover, one thing is to note facts linked to a phenotype, another is to make potential generalisations of a genetic nature. To reach a satisfactory answer, we must focus on long timeframes and look at the *great* technological innovations to which significant results in terms of experimental verification can be attached. This is the classic case of the invention of writing (McLuhan 1962; Ong 1982; Antinucci 2011; D'Errico and Colagè 2018; Bruner et al. 2018), perhaps extendible to the writing-image integration allowed today by specific digital-technology resources (Montani 2018).

The first articulation along which a specific contribution of the arts can appear emerges against this extremely problematical backdrop. Art has found itself explicitly interpreting the idea of a fundamental techno-aesthetic determinism, bringing into play the aptitude for *free and openminded experimentation* that modernity granted and has never really revoked. Looking simply at the years following the seismic happening of the technical production of images (photography and films), it is very clear that one of the purposes explicitly pursued by the first avantgardes was that of experimentally exploring previously unseen sensitive affordances, sometimes pushing them to the extremes (see among others Benjamin 1969; Eisenstein 2002; Moholy-Nagy 2018; Vertov 1992).

The second line of articulation, which I have called "revelatory", is closely coordinated with the first and, in part, constitutes one of its internal specifications. The point being that the historically framed human being is so closely engaged in his/her "natural" technical dimension that he/she is unable to distance himself/herself far enough to acquire an objective vision of what are, each time, the demands that this condition is putting forward and in what position it has placed himself/herself. To make just one example linked to a lively current debate (Floridi et al. 2018; Floridi 2019), what "positioning" has the human being in the interaction with the latest generation of artificial intelligence algorithms powered by machine learning techniques? Has the human being perhaps become a mere component—e.g. a provider of sensitivity—in an individuation process (Simondon 1989) that no longer concerns him/her *individually* because the individuation applies, instead, to the *interactive whole* of which he/she is a part (Stiegler 2016)? Art exercises the function I have called "revelatory" whenever it feels pressed to respond to this kind of question in the manner it finds most congenial: repositioning the human being's sensitivity at the proper techno-aesthetic level and showing how it is changing (as in the extreme example just described). As an example we can refer to the radical rethinking of the values of truth (or the "realism" if you wish) of the images brought into play by the arts in the last 30 years, in conjunction with the rise of digital technologies. In this vast exploration of the new forms of realism, we have seen that our sensitivity for what is "real" has now assumed a "trans-media" or "inter-media" dimension that is inseparable from the reflective assumption of the specific technologies with which the single mediums process images of the world-environment (Montani 2010; Ferraris 2017).

Under the "deconstructive" heading, finally, the third line of articulation of the relationship between techno-aesthetics and the historical art phenomenon highlights a *single profile of pertinence* in the sphere of the "negativity" that has often been identified in the modern experience of the arts (Adorno 1997; Debord 2013). From this standpoint, negativity has asserted itself in two different permutations: on the one hand, as a more or less radical calling into question of the coercive power attributed to the main technological devices (and today we must, once again, think of the digital world) and as a highlighting of their internal malfunctions; on the other, as an "expropriation" of the unseen potential they hold in reserve, as not envisaged by their constituent project (Krauss 2000). In both cases, albeit to differing degrees, it is significant that, in this approach, the dialectic of the two aspects of pleasure and displeasure, or affirmativeness and negativity if you prefer, can be seen as a way of enforcing a certain human "ethical continuity" against the backdrop of the discontinuities present in the historically variable destiny of its constituent technical assignment.

The three articulations I have identified here refer to art in the aesthetic sense (Garroni 1992; Rancière 2000), but the case history would be far more extensive were we to shift the focus of the argument from the (specific) issue of art to the (far more general) one of technical creativity (Montani 2017). We can justifiably predict that this shift will become ever more pressing in the coming years.

References

Adorno, T. W. (1997). *Aesthetic theory (1970)*. London: The Athlone Press.
Antinucci, F. (2011). *Parola e immagine. Storia di due tecnologie*. Roma-Bari: Laterza.
Bateson, G. (1973). *Steps to an ecology of mind*. San Francisco: Chandler Pub. Co.
Benjamin, W. (1969). *The Work of Art in the Age of Mechanical Reproduction (1935)*. New York: Shocken Books.
Bruner, E., Fedato, A. P., Silva-Gago, M., Alonso-Alcalde, R., Tarradillos-Bernal, M., Fernandez-Durante, M. A., & Martin-Guerra, E. (2018). Chapter 12 cognitive archaeology, body cognition, and hand-tool interaction. In *Progress in brain research* (Vol. 238, pp. 325–345). London: Springer.
D'Errico, F., & Colagè, I. (2018). *Cultural exaptation and cultural neural reuse. A mechanism for the emergence of modern culture and behavior. Biological Theory, 13*, 213–227. https://doi.org/10.1007/s13752-018-0306-x.
Debord, G. (2013). *Society of the spectacle (1967)*. Detroit: Black and Red.
Eisenstein, S. M. (2002). *Metod (1943)*. Moskva: Muzej Kino, Ėjzenštejn-Centr.
Fauconnier, G. (1997). *Mappings in thought and language*. Cambridge: Cambridge University Press.
Fauconnier, G., & Turner, M. (2002). *The way we think. Conceptual blending and the mind's hidden complexities*. New York: Basic Books.
Ferraris, M. (2017). *Fare la verità, "Costellazioni"* (Vol. 4, pp. 37–48).
Floridi, L., et al. (2018). AI4People—An Ethical Framework for a Good AI Society: Opportunities, Risks, Principles, and Recommendations. Minds and Machines *28* (4):689–707.
Floridi, L. (2019). What the near future of artificial intelligence could be. In *Philosophy & technology*. London: Springer. https://doi.org/10.1007/s13347-019-00345-y.
Freud, S. (1990). *Beyond the pleasure principle (1920)*. New York: W.W. Norton & Company.
Gallagher, S. (2005). *How the body shapes the mind*. Oxford: Oxford University Press.
Gallagher, S. (2017). *Enactivist interventions. Rethinking the mind*. Oxford: Oxford University Press.
Gallese, V., & Lakoff, G. (2005). The brain's concepts. The role of the sensory-motor system in reason and language. *Cognitive Neuropsychology, 22*, 455–479.
Garroni, E. (1986). *Senso e paradosso*. Roma-Bari: Laterza.
Garroni, E. (1992). *Estetica. Uno sguardo-attraverso*. Milano: Garzanti.
Garroni, E. (2005). *Immagine, linguaggio, figura*. Roma-Bari: Laterza.
Heidegger, M. (2010). *Being and time (1927)*. New York: State University of New York Press.
Ihde, D., & Malafouris, L. (2018). *Homo faber revisited. Postphenomenology and material engagement theory. Philosophy & Technology, 32*(2), 195–214. https://doi.org/10.1007/s13347-018-0321-7.
Kant, I. (1998). *Critique of pure reason (1781)*. Cambridge: Cambridge University Press.
Kant, I. (2000a). *Critique of the power of judgment (1790)*. Cambridge: Cambridge University Press.
Kant, I. (2000b). *First Introduction to the Critique of the Power of Judgment* [1790] (pp. 1–15). Cambridge: Cambridge University Press.

Krauss, R. (2000). *A voyage on the North Sea. Art in the age of the post-medium condition*. London: Thames & Hudson.
Lakoff, G., & Johnson, M. (1980). *Metaphors we live by*. Chicago: University of Chicago Press.
Malabou, C. (2016). *Before tomorrow. Epigenesis and rationality*. Cambridge: Polity Press.
Malafouris, L. (2013). *How things shape the mind. A material engagement theory*. Cambridge: The MIT Press.
McLuhan, M. (1962). *The Gutenberg galaxy. The making of typographic man*. Toronto: University of Toronto Press.
Moholy-Nagy, L. (2018). *Painting, photography, film (1927)*. München: Bauhausbücher.
Montani, O. (2018). *Sensibilità, immaginazione e linguaggio. Processi di interiorizzazione e cultura digitale. "Bollettino della Società filosofica Italiana", Settembre-Dicembre* (pp. 25–41).
Montani, P. (2007). *Bioestetica. Senso comune, tecnica e arte nell'età della globalizzazione*. Roma: Carocci.
Montani, P. (2010). *L'immaginazione intermediale*. Roma-Bari: Laterza.
Montani, P. (2014). *Tecnologie della sensibilità*. Milano: Cortina.
Montani, P. (2017). *Tre forme di creatività. Tecnica, arte, politica*. Napoli: Cronopio.
Nanay, B. (2016). *Aesthetics as philosophy of perception*. Oxford: Oxford University Press.
Ong, W. J. (1982). *Orality and literacy. The technologizing of the world*. London: Methuen.
Panksepp, J., & Biven, L. (2012). *The archaeology of mind. Neuroevolutionary origins of human emotion*. New York: W.W. Norton & Company.
Parisi, F. (2019). *La tecnologia che siamo*. Torino: Codice.
Rancière, J. (2000). *Le partage du sensible*. Paris: La fabrique.
Ricoeur, P. (1975). *La métaphore vive*. Paris: Seuil.
Simondon, G. (1989). *L'individuation psychique et collective*. Paris: Aubier.
Simondon, G. (2014). *Sur la technique*. Paris: PUF.
Stiegler, B. (2016). *Automatic society: The future of work* (Vol. 1). Cambridge: Polity Press.
Tomasello, M. (1999). *The cultural origins of human cognition*. Cambridge: Harvard University Press.
Turner, M. (1996). *The literary mind*. Oxford: Oxford University Press.
Vertov, D. (1992). *Kino-eye. The writings of Dziga Vertov*. Berkeley: University of California Press.

The manufacturer's authorised representative in the EU is Springer Nature Customer Service Centre GmbH, Europaplatz 3, 69115 Heidelberg, Germany. If you have any concerns regarding our products, please contact ProductSafety@springernature.com

Printed and bound by CPI Group (UK) Ltd, Croydon, CR0 4YY
30/09/2026
02225453-0014